Multiple Choice Questions in Anatomy and Neurobiology for Undergraduates

Multiple Choice Questions in Anatomy and Neurobiology for Undergraduates

Michael J. Blunt, MB, BS(Lond.), PhD(Lond.), FRACS(Hon.)
Challis Professor of Anatomy,
The University of Sydney;
Staff Associate, Regional Teacher Training Centre for
Health Personnel, The University of New South Wales, Australia

M. Girgis, MD(Khart.), MA(Camb.), PhD(Lond.), MRANZCP.
Senior Lecturer in Anatomy,
The University of Sydney, Australia

BUTTERWORTHS
LONDON-BOSTON
Sydney-Wellington-Durban-Toronto

The Butterworth Group

United Kingdom	**Butterworth & Co (Publishers) Ltd** London: 88 Kingsway, WC2B 6AB
Australia	**Butterworths Pty Ltd** Sydney: 586 Pacific Highway, Chatswood, NSW 2067 Also at Melbourne, Brisbane, Adelaide and Perth
Canada	**Butterworth & Co (Canada) Ltd** Toronto: 2265 Midland Avenue, Scarborough, Ontario, M1P 4S1
New Zealand	**Butterworths of New Zealand Ltd** Wellington: T & W Young Building, 77–85 Customhouse Quay, 1, CPO Box 472
South Africa	**Butterworth & Co (South Africa) (Pty) Ltd** Durban: 152–154 Gale Street
USA	**Butterworth (Publishers) Inc** Boston: 19 Cummings Park, Woburn, Mass. 01801

First published 1979

© Butterworth & Co (Publishers) Ltd, 1979

ISBN 0 407 00153 0

British Library Cataloguing in Publication Data

Blunt, Michael J
    Multiple choice questions in anatomy and
    neurobiology for undergraduates.
    1. Anatomy, Human – Problems, exercises, etc.
    2. Neurobiology – Problems exercises, etc
    I. Title II. Girgis, M
    611'.0076   QM32   78-40790

    ISBN 0-407-00153-0

Typeset by Butterworths Litho Preparation Department
Printed and bound by Billings & Sons Ltd, Guildford & London

Preface

The preparation of reliable, unambiguous multiple choice questions constitutes a lengthy and exacting task. Even though a committee of individuals may have scrutinized questions before their acceptance into a bank, ambiguities frequently become apparent only after use and in consequence of item analysis of question data. For this reason academic departments tend to hoard multiple choice questions, aspiring, usually unsuccessfully, to keep them secret from the students on whom they are used. Opportunities for learning from the questions are thus hindered rather than fostered.

It is hoped that the publication of this collection of questions will somewhat diminish the need for such hoarding and secrecy. The majority are accompanied by item analyses which would enable them to be used with confidence for testing relevant subject content.

For students, especially those subjected to objective tests of performance, the opportunity to obtain feed-back from responses to multiple choice questions is always welcome. In this instance the feed-back will be certainly fuller and more reliable than may be obtained from previously untested questions.

This collection of questions has been drawn from a question bank built up for specifically undergraduate tests through several years of use. Many past and present members of the academic staff of the Department of Anatomy in the University of Sydney have been involved in either its initial compilation or subsequent modifications. To all of these we extend our grateful thanks. In particular we are indebted to Dr. A. E. Sefton and Professor W. Burke for the physiology questions in the section of neurobiology, and to Dr. M. Arnold for carefully checking the manuscript and proofs.

Finally we wish to acknowledge the painstaking typing and the useful suggestions on format which emanated from Ms. Robyn Vears, Departmental Secretary and Miss Maria Karlsson-Lillas, former occupant of that position.

Sydney

Michael Blunt
Makram Girgis

Contents

Chapter 1

Description, Aims,
Suggestions to Users

This book of multiple choice questions is primarily intended for the use of medical undergraduates, for whom it has three main aims.

(1) To facilitate the learning of topographical anatomy and neurobiology.
(2) To provide a measure by which progress may be assessed.
(3) To provide practice in anatomy MCQ's.

It is, in addition, hoped that the questions presented will be useful to academic anatomists building up their own question banks.

Questions have been grouped for each visceral system and the nervous system; in the case of the body framework (including peripheral nerves and blood vessels), however, a regional grouping has been adopted. In pursuit of the aim to facilitate learning all headings and sub-headings follow the order used in *A New Approach to Teaching and Learning Anatomy* (Blunt, 1976). Moreover, each question is referred to an objective or series of objectives in that book. Thus, individuals using *A New Approach to Teaching and Learning Anatomy* as a companion volume may readily obtain immediate feed-back on their achievement of the objectives which it designates. In most instances reference is made to one or more general instructional objectives and to one or more specific learning objectives. On occasions, however, when it is considered that there is no sufficiently close match with specific learning objectives, only a general instructional objective is indicated.

In fostering the aim of providing students with an immediate measure of progress each question has been accompanied alongside by its correct answer. Most questions are also accompanied by an indication of the percentage of students who obtained the correct answer on occasions of previous settings and by a biserial correlation coefficient (r biserial), indicative of the capacity of the question to distinguish between more able and less able students. Inclusion also indicates the statistical significance of the r biserial at the 0.05

1

level of probability. Thus, the student can tell at a glance whether the question he has attempted has been validated by previous use, and, if so, what levels of difficulty and discrimination it presented. The figures shown relate to groups of about 250 undergraduate medical students. In instances where questions have been set several times, there may be substantial variation in performance reflecting essentially differences in preparedness on the basis of teaching given. Clearly no absolute significance is attached to the performance levels recorded, but they serve as a general guide to levels of difficulty experienced by a group of students who had prepared themselves for a test. Biserial correlation coefficients vary from 0 to 1 and may be given the sign + or − according to the direction of the discrimination. All those given in fact represent positive discrimination so the sign has been omitted. The closer the values shown approach to 1 the better the questions distinguish between students getting higher total marks and less able colleagues. The indication that the r biserial is significant at a probability level of 0.05 means that a value as large as this will occur by chance only 5 times in 100 samples.

Questions which have been tried but found wanting have not been included in this collection. Thus, an item analysis, when given, is indicative of valid testing in relation to the objectives on which the question is based. By using the data provided instructors may be assisted in the preparation of criterion referenced test papers at predetermined levels of difficulty. Student users should appreciate that with the higher r biserials there is a greater likelihood that more able members of previous cohorts of students sitting a question did not encounter ambiguities or semantic problems. Such problems are a well known hazard of multiple choice questions and the best guarantee that they have been overcome is the evidence of an r biserial reaching at least the 0.05 level of significance. In the instances in which no item analysis is given, questions have not been tested by previous use and their reliability is as yet an unknown quantity.

Each chapter of the book contains three types of questions, designed, according to the Hubbard and Clemans (1961) classification, to test knowledge in different ways. Type A items involve choosing a single correct answer from five available choices. Type E items are used to elicit information on the cause—effect relationship between sets of data. Type K items call for the perception of one or more correct responses among four alternatives and the responses may be grouped in five different ways. Type A items thus predominantly require the recall of isolated items of knowledge, whereas Type E and K questions additionally require perception of the congruence or cohesiveness between sets of data (Blunt and Blizard, 1975).

Suggestions to users

(1) Use a strip of paper or a ruler to cover the right-hand column of data which includes the correct answers to questions.

(2) Determine how many questions will be attempted and set a total time limit of 1¼ minutes per question.

(3) Jot down your answers to questions, remembering that if you do *not* commit yourself to paper it is easy enough to delude yourself as to what you *would* have decided!

(4) Check whether your scores compare well with those indicated in the data column.

(5) Analyse the reasons for incorrect answers. If the reason is simply insufficient knowledge the remedy is obvious. If, on the other hand, your incorrect response was associated with the feeling that you knew the correct answer, it may have been due to insufficiently accurate reading of the question or to incorrect interpretation of it. In the latter case perusal of the r biserial will provide some indication of the likelihood of ambiguity.

References

Blunt, M. J. (1976). *A New Approach to Teaching and Learning Anatomy: Objectives and Learning Activities.* London: Butterworths

Blunt, M. J. and Blizard, P. J. (1975). Recall and retrieval of anatomical knowledge. *Br. J. med. Educ.* 9, 255

Hubbard, P. J. and Clemans, W. V. (1961). *Multiple Choice Examinations in Medicine. A Guide for Examiner and Examinee.* Philadelphia: Lea and Febiger

Chapter 2

The Upper Limb

Type A items (Hubbard and Clemans, 1961) (Questions 1–88)

These involve choosing one correct answer from five available choices. The instruction for these items is as follows:

> *Each of the incomplete statements below is followed by five suggested answers or completions. Select the one which is best in each case.*
>
> (1)* (2)*(3)*

A(i) Shoulder region (Questions 1–22)

1   The serratus anterior muscle

    (A) is a medial (downward) rotator of the scapula.
    (B) is a lateral (upward) rotator of the scapula.
    (C) is a retractor (adductor) of the scapula.
    (D) acts in association with the subscapularis in rotation of the scapula.
    (E) participates in adduction of the shoulder.

Objectives†   7; I; 2; (2.2).

    0.40   46   B
    0.65   79

---

* (1) Biserial coefficient
   (2) Percentage of correct answers   }   *See* pages 1–3 for explanation
   (3) Correct answer

† *See* pages 1–3 for explanation

2   Upward (lateral) rotation of the scapula is mainly
     produced by the

(A) trapezius and rhomboid muscles.
(B) trapezius and serratus anterior muscles.
(C) serratus anterior and levator scapulae muscles.
(D) levator scapulae and trapezius muscles.
(E) serratus anterior and teres major muscles.

Objectives   7; 1; 2; (2.3), (2.4).   *0.43*   67   **B**
                                            0.56   65

3   To test for trapezius muscle paralysis, you would ask
     the patient to

(A) abduct the arm fully.
(B) flex the arm fully.
(C) push against a wall.
(D) shrug the shoulder.
(E) adduct the arm against resistance.

Objectives   7; 1; 2; (2.2).   *0.40*   92   **D**

4   Features of the scapula include

(A) coracoid process continuous with spine of
    scapula.
(B) acromion process the most laterally projecting
    part.
(C) coracoid process projecting forwards and medially.
(D) glenoid fossa projecting from the spine.
(E) glenoid fossa projecting from coracoid process.

Objectives   7; 1; 1; (1.2).   7; 1; 3; (3.1).   *0.32*   81   **B**
                                                    0.35   88

5   In abduction of the arm

(A) the clavicle remains fixed.
(B) the scapula retracts (adducts).

(C) scapular movement is at first more rapid than
     movement of the humerus.
(D) the scapula rotates medially (downwards).
(E) the medial end of the clavicle moves downwards
     on the articular disc.

Objectives   7; I; 1; (1.7), (1.10).   7; I; 3; (3.6).   *0.53*   76   E

6   The intracapsular structures of the shoulder joint
      include the tendon of the

      (A) pectoralis major muscle.
      (B) subscapularis muscle.
      (C) supraspinatus muscle.
      (D) long head of biceps muscle.
      (E) short head of biceps muscle.

      Objectives   7; I; 1.   *0.21*   62   D

7   The sternoclavicular joint

      (A) has two separate joint cavities.
      (B) lies at the level of the second costal cartilage.
      (C) is strengthened by the conoid ligament.
      (D) is a fibrous joint.
      (E) has none of the above properties.

      Objectives   7; I; 1; (1.7).   *0.48*   79   A

8   The close-packed position of the shoulder joint occurs
      with the arm

      (A) by the side of the body.
      (B) at 90° abduction with medial rotation.
      (C) at 90° abduction with lateral rotation.
      (D) at 180° abduction with medial rotation.
      (E) at 180° abduction with lateral rotation.

      Objectives   7; I; 1.   *0.28*   15   E

9   The muscle pair most important in abduction at the
     gleno-humeral joint is

     (A) deltoid and subscapularis.
     (B) deltoid and supraspinatus.
     (C) supraspinatus and subscapularis.
     (D) teres major and subscapularis.
     (E) deltoid and teres major.

Objectives   7; 1; 2; (2.3).   *0.22*   97   **B**
                                                      0.25   97
                                                      0.64   95

10   The rotator cuff muscles of the shoulder

     (A) are supplied by the radial and suprascapular
         nerves.
     (B) have important attachments to the capsule of
         the shoulder joint.
     (C) include the teres major muscle.
     (D) are the only muscles involved in lateral rotation.
     (E) are all adductors of the arm at the shoulder.

Objectives   7; 1; 2.   *0.34*   74   **B**
                                                      0.46   68

11   Adduction at the gleno-humeral joint is produced by
     the

     (A) pectoralis minor.
     (B) deltoid.
     (C) supraspinatus.
     (D) subclavius.
     (E) pectoralis major.

Objectives   7; 1; 2; (2.2), (2.3).   –   –   **E**

12   In a normal antero-posterior radiograph of the shoulder
     region, taken in the anatomical position, the most
     lateral bony feature is

     (A) the acromion.
     (B) the lesser tubercle.

(C) the coracoid process.
(D) the greater tubercle.
(E) none of the above.

Objectives   7; I; 1; (1.8).   *0.36*   72   D

13   The humerus has

(A) a greater tubercle located medial to the lesser
       tubercle.
(B) the capsule of the shoulder joint attached around
       its surgical neck.
(C) a capitulum which articulates with the olecranon
       process.
(D) a covering of synovial membrane over its head.
(E) a greater tubercle which projects further laterally
       than the acromion.

Objectives   7; I; 1; (1.4), (1.5), (1.6).
                   7; I; 3; (3.1), (3.4).   *0.40*   68   E
                                                                 0.44   82
                                                                 0.47   85

14   The clavicle

(A) articulates by a synovial joint with the coracoid
       process.
(B) underlies the coracoid process.
(C) articulates (via an articular disc) with the
       manubrium sterni and first costal cartilage.
(D) articulates supero-laterally with the acromion
       process.
(E) presents a conoid tubercle on the inferior surface
       of the shaft close to the medial end.

Objectives   7; I; 1; (1.1), (1.2), (1.3), (1.7).   *0.31*   41   C

15   Medial rotation at the gleno-humeral joint may be
      performed by the

(A) subscapularis.
(B) supraspinatus.

(C) infraspinatus.
(D) teres minor.
(E) posterior fibres of deltoid.

Objectives 7; l; 2; (2.3).   *0.54* 94   A

16 In the anatomical position, the

  (A) vertebral margin of the scapula diverges from the
      sagittal plane.
  (B) coracoid process points medially.
  (C) inferior angle of the scapula overlies the 6th rib.
  (D) acromion occupies a horizontal plane.
  (E) above statements are not correct.

Objectives 7; l; 1; (1.2).   7; l; 3; (3.1), (3.2), (3.7). *0.30* 33   D

17 Abduction at the shoulder in the scapular plane to
    180° requires

  (A) a laterally rotated gleno-humeral joint.
  (B) contraction of the rotator cuff muscles.
  (C) retraction, elevation and upward rotation of the
      scapula.
  (D) more scapular rotation than gleno-humeral
      movement.
  (E) scapular movement followed by gleno-humeral
      movement.

Objectives 7; l; 2; (2.3).   7; l; 3; (3.6).   7; l; 3.   —   —   A

18 The latissimus dorsi muscle

  (A) is a lateral rotator of the humerus.
  (B) is a lateral (upward) rotator of the scapula.
  (C) lies wholly inferior to the scapula.
  (D) is an adductor and extensor at the gleno-humeral
      joint.
  (E) is an extensor and lateral rotator at the gleno-
      humeral joint.

Objectives 7; l; 2; (2.2).   *0.43* 90   D

19   If the trapezius is paralysed

(A) elevation of the shoulder is impaired.
(B) medial rotation of the arm is impaired.
(C) lateral rotation of the arm is impaired.
(D) the scapula cannot be protracted.
(E) medial (downward) rotation of the scapula is
     impaired.

Objectives   7; l; 2; (2.2).   0.37   91   A

20   A muscle not included among the medial rotators of
     the shoulder joint is

(A) pectoralis major.
(B) latissimus dorsi.
(C) teres major.
(D) teres minor.
(E) subscapularis.

Objectives   7; l; 2; (2.3).   0.37   96   D

21   The muscle pair which most importantly assists in
     elevating the arm above the head is

(A) trapezius and pectoralis minor.
(B) levator scapulae and serratus anterior.
(C) rhomboid major and serratus anterior.
(D) rhomboid major and levator scapulae.
(E) trapezius and serratus anterior.

Objectives   7; l; 2; (2.3).   −   −   E

22   Protractors of the scapula include the

(A) pectoralis major.
(B) subscapularis.
(C) serratus anterior.
(D) trapezius.
(E) rhomboid muscles.

Objectives   7; l; 2; (2.3).   0.52   73   C

A(ii) Shoulder region including nerve supply (Questions 23–33)

23   The deltoid muscle

   (A) is supplied by the musculocutaneous nerve.
   (B) acts as both a flexor and an extensor of the arm.
   (C) is an adductor of the arm.
   (D) acts unaided in abduction of the arm.
   (E) is a depressor of the scapula.

   Objectives   7; 1; 2; (2.2).   *0.44*   81   B

24   The infraspinatus muscle

   (A) is supplied by the suprascapular nerve.
   (B) is supplied by the axillary nerve.
   (C) is a medial rotator of the humerus.
   (D) is supplied by the subscapular nerve.
   (E) has none of the above properties.

   Objectives   7; 1; 2; (2.2), (2.3).   7; IV; 1.   *0.50*   84   A

25   The serratus anterior muscle

   (A) is supplied by the thoraco-dorsal nerve.
   (B) is a retractor (adductor) of the scapula.
   (C) helps in abduction of the arm.
   (D) rotates the scapula medially (downwards).
   (E) has none of the above properties.

   Objectives   7; 1; 2; (2.2), (2.3).   7; IV; 1.   *0.38*   66   C

26   The teres major muscle

   (A) is supplied by the radial nerve.
   (B) runs parallel to and above the teres minor.
   (C) obtains its nerve supply from the axillary nerve.
   (D) assists in adduction and medial rotation of the
        humerus.
   (E) has none of the above properties.

   Objectives   7; 1; 2; (2.2), (2.3).   7; IV; 1; (1.5).   *0.48*   92   D

27   The deltoid muscle

   (A) acts unaided in abduction of the arm.
   (B) is a depressor of the scapula.
   (C) is an adductor of the arm against resistance.
   (D) acts in abduction in association with
        supraspinatus.
   (E) is supplied by the musculo-cutaneous nerve.

   Objectives   7; I; 2; (2.2), (2.3).   7; IV; 1; (1.5).   *0.32*   99   D

28   The serratus anterior muscle

   (A) is supplied by the thoraco-dorsal nerve.
   (B) is a retractor of the scapula.
   (C) helps in abduction of the arm.
   (D) interdigitates with eight slips of the latissimus
        dorsi.
   (E) has none of the above properties.

   Objectives   7; I; 2.   7; IV; 1.   *0.38*   40   C
                                                                  0.46   30

29   Lateral rotation of the arm at the gleno-humeral
      joint is

   (A) produced by contraction of teres major muscle.
   (B) associated with adduction of the arm at the
        gleno-humeral joint.
   (C) produced by contraction of muscles supplied by
        the 5th and 6th cervical spinal nerves.
   (D) produced by contraction of subclavius muscle.
   (E) produced by contraction of subscapularis muscle.

   Objectives   7; I; 2; (2.3).   7; IV; 1; (1.6).   *0.55*   71   C
                                                                  0.55   72

30   The coracobrachialis muscle

   (A) is supplied by the median nerve.
   (B) is supplied by the axillary nerve.

(C) is an abductor of the humerus.
(D) is a flexor of the elbow.
(E) has none of the above properties.

Objectives   7; I; 2; (2.1), (2.2).   7; IV; 1; (1.5).   *0.45*   88   E

31   The supinator muscle

(A) laterally rotates the radius around the ulna.
(B) is the only muscle producing supination.
(C) may produce flexion at the elbow.
(D) medially rotates the ulna around the radius.
(E) is supplied by the median nerve.

Objectives   7; III; 1; (1.5), (1.6).   7; IV; 1; (1.2).   –   –   A

32   The trapezius muscle

(A) is inactive in elevation of the arm.
(B) has the same nerve supply as the teres major.
(C) aids in lateral rotation at the gleno-humeral joint.
(D) is paralysed in an upper trunk brachial plexus
        lesion.
(E) acts as a mechanical couple in lateral (upward)
        rotation of the scapula.

Objectives   7; I; 2; (2.2).   7; IV; 1.   *0.47*   82   E

33   The long head of the biceps brachii muscle

(A) lies partly within the shoulder joint.
(B) runs superficial to the transverse ligament.
(C) adducts the shoulder joint.
(D) is innervated indirectly from the medial cord of
        the brachial plexus.
(E) has none of the above properties.

Objectives   7; I; 1; (1.6).   7; I; 2; (2.2).
                7; IV; 1; (1.1).   –   –   A

A(iii) Elbow region (Questions 34–38)

34   In relation to the radius and ulna

(A) the bones are of equal length.
(B) both participate in the radio-carpal joint.
(C) the heads of both bones rotate in supination.
(D) the radius articulates with the trochlea of the
humerus.
(E) the radius is the principal weight or force
transmitting bone of the forearm.

Objectives   7; I; 1; (1.1).
7; III; 1; (1.1), (1.2), (1.3), (1.6).   0.27   66   E

35   The apex of the cubital fossa is defined by

(A) biceps brachii and pronator teres.
(B) brachioradialis and pronator teres.
(C) biceps brachii and supinator.
(D) brachioradialis and supinator.
(E) none of the above.

Objectives   7; II; 1.   0.38   54   B
                                                0.49   51

36   At the lower end of the humerus

(A) maximum growth in length takes place.
(B) epiphysial plates persist in the adult as thin
layers of hyaline cartilage.
(C) there are numerous separate epiphyses which
remain unfused.
(D) epiphysial plates show up as dark bands on
radiographs
(E) none of the above situations obtain.

Objectives   7; II; 1; (1.7), (1.8).   0.28   21   D

37 The medial epicondyle of the humerus

(A) is the common extensor origin of the muscles
of the forearm.
(B) lies anterior to the radial nerve.
(C) lies immediately lateral to the capitulum.
(D) becomes palpable on pronation.
(E) has none of the above properties.

Objectives 7; II; 1; (1.1), (1.8). 7; IV; 1; (1.4). *0.36* 75 E

38 The triceps muscle

(A) acts at the elbow but not at the shoulder joint.
(B) is the only extensor at the elbow.
(C) is supplied by the 5th cervical segment of the
spinal cord.
(D) acts mainly through its medial head in unresisted
extension of the elbow.
(E) is partly supplied by the axillary nerve.

Objectives 7; II; 1; (1.1), (1.13).
7; IV; 1; (1.3), (1.5), (1.6). *0.29* 55 D
0.38 54

A(iv) Forearm, wrist and hand (Questions 39–44)

39 The tendons of the flexor digitorum profundus muscle

(A) lie deep to the tendons of flexor digitorum
superficialis.
(B) divide to accommodate the tendons of the flexor
digitorum superficialis.
(C) are enclosed within synovial sheaths separate from
those of the flexor digitorum superficialis.
(D) separate from one another distal to the flexor
retinaculum.
(E) give attachment to lumbrical muscles on their
medial sides.

Objectives 7; III; 3; (3.1), (3.4). *0.40* 47 D

40   The wrist joint (radio-carpal joint)

(A) has a synovial cavity continuous with that of the
inferior radio-ulnar joint.
(B) has a synovial cavity continuous with that of the
midcarpal joint.
(C) permits a considerable amount of flexion,
extension and adduction, but little abduction.
(D) has the distal (articular) surface of the radius
facing distally, medially and dorsally.
(E) has the flexor retinaculum anterior to it.

Objectives   7; III; 2.   0.27   56   C
                                                0.32   53

41   The interossei and lumbrical muscles

(A) flex at the interphalangeal and extend at the
metacarpophalangeal joints.
(B) flex at both interphalangeal and metacarpophalangeal joints.
(C) flex at the distal interphalangeal and extend at
the proximal interphalangeal joints.
(D) extend at the interphalangeal and flex at the
metacarpophalangeal joints.
(E) extend at both the interphalangeal and metacarpophalangeal joints.

Objectives   7; III; 3; (3.5).   0.69   71   D
                                                0.66   82

42   Power grip involves

(A) abduction of the thumb.
(B) palmar flexion at the wrist.
(C) opposition of the thumb.
(D) radial deviation at the wrist.
(E) none of the above.

Objectives   7; III; 3; (3.7).   0.39   60   E
                                                0.23   68

43   The middle finger

(A) is adducted by the second palmar interosseous
    muscle.
(B) is flexed by the third lumbrical muscle.
(C) is adducted by the third palmar interosseous
    muscle.
(D) is abducted by the second dorsal interosseous
    muscle.
(E) has none of the above properties.

Objectives   7; III; 3; (3.5).   0.64   91   D
                                           0.41   74
                                           0.43   76

44   The distal palmar crease

(A) marks the distal margin of the flexor retinaculum.
(B) is proximal to the surface marking of the super-
    ficial palmar arch.
(C) is opposite the heads of the metacarpals.
(D) is produced by movements of the fingers.
(E) marks the distal limit of the radial bursa.

Objectives   7; III; 3; (3.1), (3.3), (3.8).
              7; IV; 2; (2.2).   0.26   38   D

A(v) Forearm, wrist and hand, including nerve supply (Questions 45–59)

45   The carpal canal (tunnel) contains

(A) the flexor carpi ulnaris tendon.
(B) the ulnar artery.
(C) the radial artery.
(D) the deep branch of the ulnar nerve.
(E) none of the above.

Objectives   7; III; 3; (3.1).   7; IV; 1; (1.2).
              7; IV; 2; (2.1).   0.40   54   E

46   The abductor pollicis longus muscle

(A) is supplied by the radial nerve.
(B) is attached to the head of the first metacarpal.
(C) lies in the front of the forearm.
(D) extends the wrist.
(E) passes superficial to the extensor retinaculum.

Objectives   7; III; 3; (3.1).   *0.56*   81   A
                                                             0.49   75

47   The extensor carpi radialis longus muscle

(A) is supplied by the median nerve.
(B) is the sole radial extensor of the wrist.
(C) extends at the metacarpophalangeal joint of the
    thumb.
(D) extends at the carpometacarpal joint of the
    thumb.
(E) has none of the above properties.

Objectives   7; III; 2; (2.5).   7; III; 3; (3.1)
             7; IV 1; (1.5).   –   –   E

48   Pronation

(A) is dependent upon an intact ulnar nerve.
(B) is a stronger movement than supination.
(C) may be produced by extensor carpi radialis longus.
(D) is dependent upon the integrity of C6.
(E) is dependent upon an intact radial nerve.

Objectives   7; III; 1; (1.5).   7; IV; 1; (1.5).   *0.67*   79   D

49   The reflex elicited by tapping the biceps brachii
      tendon

(A) tests for integrity of spinal segments C6 and C7.
(B) is an example of a 3 neurone reflex arc.
(C) is lost following a complete lesion of the median
    nerve.

(D) is completely lost if biceps brachii is paralysed.
(E) involves contraction of all the flexors of the
     elbow.

Objectives   7; II; 1; (1.13).
             7; IV; 1; (1.3), (1.5), (1.6).   *0.35*   51   D

50   The flexor digitorum superficialis muscle

(A) has a deep part which ends in tendons to the
     middle and ring fingers.
(B) has tendons completely enclosed in synovial
     sheaths.
(C) has tendons which insert into the middle
     phalanges.
(D) is innervated by both median and ulnar nerves.
(E) has none of the above properties.

Objectives   7; III; 3; (3.1).   7; IV; 1; (1.5).   *0.37*   83   C

51   Abduction of the ring finger

(A) is produced by the contraction of the 3rd dorsal
     interosseous.
(B) is produced by contraction of the 4th palmar
     interosseous.
(C) is lost when the median nerve is cut.
(D) is retained when the ulnar nerve is cut.
(E) has none of the above properties.

Objectives   7; III; 3; (3.5).   7; IV; 1; (1.5).   *0.35*   65   E

52   The flexor carpi radialis muscle

(A) assists in supination.
(B) lies medial to the flexor pollicis longus.
(C) has a tendon which lies lateral to the radial
     artery.
(D) is supplied by the radial nerve.
(E) has none of the above properties.

Objectives   7; III; 1; (1.6).   7; III; 2; (2.4).
             7; IV; 1; (1.5).   *0.34*   47   E

53   The pronator teres muscle

(A) inserts into the radial tuberosity.
(B) inserts into the coronoid process.
(C) is active only during pronation.
(D) receives motor fibres from the median nerve.
(E) receives motor fibres from cord segments C7
and C8.

Objectives   7; III; 1; (1.5).   7; IV; 1; (1.5), (1.6).   *0.47*   89   D

54   The palmar interossei

(A) are attached to tendons of flexor digitorum
profundus.
(B) insert into the extensor expansion.
(C) are supplied by the median nerve.
(D) adduct the 3rd digit.
(E) flex the middle phalanges.

Objectives   7; III; 3; (3.4), (3.5).   7; IV; 1; (1.5).   –   –   B

55   The biceps brachii muscle

(A) is a powerful pronator.
(B) is supplied by the median nerve.
(C) is a powerful supinator.
(D) is supplied by the radial nerve.
(E) has none of the above properties.

Objectives   7; III; 1; (1.5).   7;IV; 1; (1.5).   *0.57*   97   C

56   The flexor digitorum superficialis muscle

(A) is partly innervated by the ulnar nerve.
(B) flexes at the metacarpophalangeal and extends at
the interphalangeal joints.
(C) is innervated by the radial nerve.
(D) supinates the forearm from the midprone position.
(E) has none of the above properties.

Objectives   7; III; 3.   7; IV; 1; (1.5).   *0.58*   65   E
                                            0.74   83

57   The biceps brachii muscle

   (A) pronates the flexed forearm.
   (B) pronates the extended forearm.
   (C) is innervated by the ulnar nerve.
   (D) supinates the flexed forearm.
   (E) does none of the above.

   Objectives   7; III; 1; (1.5).   7; IV; 1; (1.5).   *0.25*   57   D
                                                         0.28   86

58   Palmar interossei

   (A) abduct the index finger.
   (B) insert into the extensor expansions of digits.
   (C) are supplied by the median nerve.
   (D) adduct the 3rd digit.
   (E) flex the middle phalanges of digits.

   Objectives   7; III; 3; (3.4), (3.5).   7; IV; 1; (1.5).   *0.65*   83   B

59   The extensor pollicis longus muscle

   (A) acts primarily at the interphalangeal joint.
   (B) acts primarily at the metacarpophalangeal joint.
   (C) accompanies the tendon of abductor pollicis
       longus.
   (D) receives a branch from the ulnar nerve.
   (E) has none of the above properties.

   Objectives   7; III; 3; (3.1).   7; IV; 1; (1.5).   –   –   A

A(vi) Nerves and blood vessels (Questions 60–88)

60   The brachial artery

   (A) may be compressed by upward pressure on the
       floor of the axilla.
   (B) is accompanied by the median and ulnar nerves
       into the cubital fossa.
   (C) terminates opposite the neck of the radius.

(D) is accompanied by a continuation of the cephalic
vein in the arm.

(E) has none of the above properties.

Objectives   7; IV; 2; (2.1).   —   —   C

61   The radial nerve is distributed to the

(A) biceps brachii.

(B) coracobrachialis.

(C) brachioradialis.

(D) deltoid.

(E) dorsal interossei.

Objectives   7; IV; 1; (1.5).   0.51   95   C

62   The upper trunk of the brachial plexus

(A) is derived from the 5th, 6th and 7th cervical
ventral rami.

(B) gives off the musculocutaneous nerve.

(C) gives off the suprascapular nerve.

(D) gives off the axillary nerve.

(E) gives off one root of the median nerve.

Objectives   7; IV; 1; (1.1).   0.54   65   C
                                                            0.42   59

63   The lateral cord of the brachial plexus

(A) gives rise to the axillary nerve.

(B) divides into the median and radial nerves.

(C) gives off the musculocutaneous nerve.

(D) is made up of the posterior divisions of the
trunks.

(E) has none of the above properties.

Objectives   7; IV; 1; (1.1).   0.42   79   C
                                                            0.71   96
                                                            0.67   91

64   The suprascapular nerve arises from the following
     parts of the brachial plexus

   (A) the upper roots, 5th and 6th ventral rami.
   (B) the upper trunk.
   (C) the anterior division of the middle trunk.
   (D) the lateral cord.
   (E) the posterior division of the middle trunk.

   Objectives   7; IV; 1; (1.1).   *0.33*   48   B

65   The musculocutaneous nerve supplies

   (A) the skin over the medial aspect of the forearm.
   (B) the teres minor muscle.
   (C) the coracobrachialis muscle.
   (D) the skin over the distal and medial aspect of the
       arm.
   (E) the teres major muscle.

   Objectives   7; IV; 1; (1.5).   *0.59*   96   C

66   The musculocutaneous nerve

   (A) supplies the brachioradialis.
   (B) supplies skin on the medial side of the arm.
   (C) is derived from the posterior cord of the brachial
       plexus.
   (D) supplies the skin of the medial side of the forearm.
   (E) has none of the above properties.

   Objectives   7; IV; 1; (1.1), (1.5).   *0.66*   81   E

67   The axillary nerve

   (A) is a terminal branch of the medial cord of the
       brachial plexus.
   (B) leaves the axilla below the teres major muscle.
   (C) supplies the subscapularis muscle.
   (D) supplies the teres minor and deltoid muscles.
   (E) has none of the above properties.

   Objectives   7; IV; 1; (1.1), (1.2), (1.5).   *0.21*   98   D
                                                    0.43   86

68   The musculocutaneous nerve

(A) supplies the triceps.
(B) normally pierces the biceps brachii.
(C) arises from the medial cord of the brachial plexus.
(D) supplies all the muscles of the anterior compartment of the arm.
(E) has the root values C7, C8, T1.

Objectives   7; IV; 1; (1.1), (1.2), (1.5).   *0.53*   68   **D**

69   The innervation of the lumbrical muscles parallels the innervation of the

(A) flexor digitorum superficialis.
(B) flexor digitorum profundus.
(C) extensor digitorum.
(D) interossei.
(E) two flexores carpi muscles.

Objectives   7; IV; 1; (1.5).   —   —   **B**

70   The skin of the index finger is supplied by

(A) ulnar and radial nerves.
(B) radial and median nerves.
(C) median and ulnar nerves.
(D) median nerve only.
(E) radial nerve only.

Objectives   7; IV; 1; (1.5).   *0.54*   81   **B**
                                                     0.54   71

71   The skin of the palm is supplied by

(A) ulnar and median nerves.
(B) radial and median nerves.
(C) radial and ulnar nerves.
(D) ulnar nerve alone.
(E) radial nerve alone.

Objectives   7; IV; 1; (1.5).   *0.36*   95   **A**

72   The muscles of the thenar eminence are mainly
      supplied by

      (A) the ulnar nerve.
      (B) the median nerve.
      (C) the radial nerve.
      (D) the posterior interosseous nerve.
      (E) none of the above.

      Objectives   7; IV; 1; (1.5).   *0.43*   82   B

73   The ulnar nerve gives branches to

      (A) the flexor pollicis longus muscle.
      (B) the flexor digitorum profundus muscle.
      (C) the opponens pollicis muscle.
      (D) the flexor carpi radialis muscle.
      (E) none of the above.

      Objectives   7; IV; 1; (1.5).   *0.75*   93   B
                                                          0.58   94

74   The ulnar nerve is distributed to

      (A) the 1st dorsal interosseous muscle.
      (B) flexor pollicis longus.
      (C) pronator quadratus.
      (D) abductor pollicis brevis.
      (E) none of the above.

      Objectives   7; IV; 1; (1.5).   *0.47*   86   A

75   The ulnar nerve supplies

      (A) the 2nd dorsal interosseous muscle.
      (B) flexor pollicis longus.
      (C) extensor carpi ulnaris.
      (D) abductor pollicis longus
      (E) none of the above.

      Objectives   7; IV; 1; (1.5).   *0.73*   93   E

76   The radial nerve

   (A) arises from the posterior cord of the brachial
       plexus.
   (B) has no cutaneous distribution.
   (C) supplies the coracobrachialis muscle.
   (D) supplies the radial but not the ulnar extensors of
       the carpus.
   (E) supplies the dorsal interossei.

   Objectives   7; IV; 1; (1.1), (1.5).   *0.73*   96   A

77   The intrinsic muscles of the hand are all supplied by the

   (A) median nerve.
   (B) 1st thoracic segment of the spinal cord.
   (C) lateral cord of the brachial plexus.
   (D) ulnar nerve.
   (E) radial nerve.

   Objectives   7; IV; 1; (1.1), (1.5), (1.6).   *0.35*   83   B

78   The median nerve supplies the following muscles in
      the hand:

   (A) all the lumbricals, opponens pollicis, abductor
       pollicis brevis, flexor pollicis brevis.
   (B) opponens pollicis, abductor pollicis brevis,
       adductor pollicis, 1st and 2nd lumbricals.
   (C) opponens pollicis, flexor pollicis brevis, abductor
       pollicis brevis, 2nd and 3rd palmar interossei.
   (D) opponens pollicis, abductor pollicis brevis, flexor
       pollicis brevis, adductor pollicis.
   (E) opponens pollicis, abductor pollicis brevis, flexor
       pollicis brevis, 1st and 2nd lumbricals.

   Objectives   7; IV; 1; (1.5).   *0.43*   96   E

79   The median nerve

   (A) comes from the posterior cord of the brachial
       plexus.
   (B) supplies the abductor pollicis brevis.

(C) passes superficial to the flexor retinaculum.
(D) supplies all the lumbricals.
(E) has none of the above properties.

Objectives   7; IV; 1; (1.1), (1.2), (1.5).   *0.26*   62   B

80   The radial nerve

(A) supplies the skin on the lateral side of the palm
of the hand.
(B) is in contact with the medial epicondyle of the
humerus.
(C) is distributed to the abductor pollicis longus
muscle.
(D) is distributed to the adductor pollicis muscle.
(E) is distributed to all the muscles concerned with
supination.

Objectives   7; IV; 1; (1.2), (1.4), (1.5).   *0.39*   56   C
                                                    0.57   76

81   The nerve most intimately related to the capsule of
the shoulder joint is the

(A) radia'.
(B) axillary.
(C) median.
(D) ulnar.
(E) musculocutaneous.

Objectives   7; IV; 1; (1.2).   *0.48*   95   B

82   The long thoracic nerve supplies

(A) the latissimus dorsi muscle.
(B) the rhomboid major and minor muscles.
(C) the serratus anterior muscle.

(D) the subscapularis muscle.
(E) none of the above.

Objectives   7; IV; 1.   *0.67*   94   C
                                                           0.46   82

83   The ulnar nerve

(A) enters the forearm through the pronator teres
    muscle.
(B) supplies the interosseous muscles.
(C) can be rolled against the lateral epicondyle of the
    humerus.
(D) supplies the pronator quadratus muscle.
(E) passes deep to the flexor retinaculum.

Objectives   7; IV; 1; (1.2), (1.4), (1.5).   —   —   B

84   The basilic vein

(A) ascends along the lateral margin of the biceps.
(B) joins the cephalic vein to form the axillary vein.
(C) joins the cephalic vein to form the median cubital
    vein.
(D) joins the brachial veins to form the axillary vein.
(E) is always connected to the cephalic vein by the
    median vein of the forearm.

Objectives   7; IV; 2; (2.6).   *0.31*   57   D
                                                           0.39   39

85   In the cubital fossa the

(A) axillary artery divides into brachial and inter-
    osseous arteries.
(B) basilic vein becomes the brachial vein.
(C) median cubital vein becomes the brachial vein.
(D) radial artery enters the dorsal compartment of
    the forearm.

(E) median cubital vein communicates with deep
     veins of the forearm.

Objectives   7; IV; 2; (2.1), (2.6).   *0.27*   66   E
                                                0.32   48

86   The median cubital vein

(A) overlies the ulnar nerve.
(B) is separated from the brachial artery by the
     bicipital aponeurosis.
(C) is essential for the venous drainage of the upper
     limb.
(D) pierces the deep fascia.
(E) is practically constant in its location.

Objectives   7; IV; 2; (2.6), (2.7).   *0.44*   90   B

87   The brachial artery

(A) commences at the upper border of the teres
     major muscle.
(B) is in direct contact with the humerus.
(C) has the biceps tendon medial to it.
(D) is readily compressible.
(E) is accompanied throughout by the basilic vein.

Objectives   7; IV; 2; (2.1), (2.2), (2.4), (2.6).   *0.52*   72   D
                                                     0.31   60

88   The cephalic vein

(A) is a continuation of the ulnar side of the dorsal
     venous network.
(B) ends in the subclavian vein.
(C) lies deep to the fascia in the cubital fossa.
(D) ends in the axillary vein.
(E) has none of the above properties.

Objectives   7; IV; 2; (2.6), (2.10).   *0.30*   73   D

Type E items (Hubbard and Clemans, 1961) (Questions 89–121)

These items are of the assertion–reason type and there are five possible responses. The instruction for these items is as follows:

Each question consists of an assertion and a reason. Responses should be chosen as follows:

(A) if the assertion and reason are true statements and the reason is a correct explanation of the assertion;

(B) if the assertion and reason are true statements but the reason is NOT a correct explanation of the assertion;

(C) if the assertion is true but the reason is a false statement;

(D) if the assertion is false but the reason is a true statement;

(E) if both assertion and reason are false statements.

E(i) Shoulder region (Questions 89–93)

89   One way to demonstrate the pectoralis major in action would be to instruct the subject to place his hand on the hip and press hard

BECAUSE

this action involves adduction of the arm against resistance.

Objectives   7; 1; 2; (2.2), (2.3).   *0.41*   87   A

90   The sternoclavicular joint has two synovial cavities

BECAUSE

it is completely divided by an articular disc.

Objectives   7; 1; 1; (1.7), (1.12).   *0.25*   74   A

91   Rupture of the supraspinatus tendon impairs abduction at the gleno-humeral joint

BECAUSE

the supraspinatus is an important synergist with the deltoid in abduction.

Objectives   7; 1; 2; (2.3).   –   –   A

92 Dislocations of the gleno-humeral joint are usually
initially in an inferior direction

BECAUSE

the capsule of the shoulder joint is unsupported by
muscle insertions inferiorly.

Objectives 7; I; 1; (1.12).   *0.27* 66   A

93 The coraco-clavicular ligament takes strain off the
acromio-clavicular joint

BECAUSE

it is concerned in weight transmission from the upper
limb to the axial skeleton.

Objectives 7; I; 1; (1.6).   *0.13* 37   A

E(ii) Elbow, wrist and hand (Questions 94–105)

94 Shifts in the axis round which pronation takes place
may be produced

BECAUSE

the ulna may be abducted by the anconeus muscle.

Objectives 7; III; 1; (1.6).   –   –   A

95 Supination is usually a more powerful movement than
pronation

BECAUSE

this movement is in part produced by contraction of
the powerful brachialis muscle.

Objectives 7; III; 1; (1.6).   *0.47* 78   C

96   The hand moves with the radius in supination

BECAUSE

proximal carpal bones articulate with the radius and
not the ulna.

Objectives   7; III; 1; (1.6).   *0.40*   78   A

97   Shifts in the effective axis round which pronation
takes place may be produced in part

BECAUSE

there may be simultaneous lateral rotation at the
shoulder joint or extension at the elbow.

Objectives   7; III; 1; (1.6).   –   –   A

98   Abduction of the ulna commonly occurs during
pronation

BECAUSE

pronation is always performed through an axis passing
through the middle finger.

Objectives   7; III; 1; (1.6).   –   –   C

99   Adduction (ulnar deviation) at the wrist joint is more
restricted than abduction (radial deviation)

BECAUSE

the radial styloid is more distal than the ulnar styloid process.

Objectives   7; III; 2; (2.6).   *0.48*   70   D
                                                                     0.54   70

100   The head of the ulna articulates with bones of the
proximal row of the carpus

BECAUSE

the wrist joint is continuous with the cavity of the

inferior radio-ulnar joint.

Objectives   7; III; 1; (1.3).   7; III; 2; (2.2).   —   —   E

101 The interossei of the hand flex at the interphalangeal
     joints

     BECAUSE

     they have insertions into the dorsal extensor
     expansions of the digits.

     Objectives   7; III; 3; (3.5).   *0.57* 90   D

102 Power grip depends upon active thenar muscles

     BECAUSE

     the thumb is functionally the most important digit.

     Objectives   7; III; 3; (3.7).   *0.32* 37   D
                                                     0.66 79
                                                     0.45 52

103 A precision grip depends largely on the movement of
     opposition of the thumb

     BECAUSE

     this movement results in a tip to tip contact between
     the thumb and the index finger.

     Objectives   7; III; 3; (3.7).   *0.30* 46   C

104 The extensor digitorum is a comparatively weak
     extensor at the interphalangeal joints of the fingers

     BECAUSE

     most of its pull is expended on the metacarpo-
     phalangeal joints.

     Objectives   7; III; 3; (3.1).   *0.36* 36   A

105 The distal transverse flexion crease of the palm
     corresponds roughly to the level of the medial two
     metacarpophalangeal joints

     BECAUSE

     it represents a line of stasis and tethering of the skin
     produced by flexion at the joints mentioned.

     Objectives   7; III; 3; (3.8).   *0.24*   13   D

E(iii) Elbow, wrist and hand, including nerve supply (Questions 106–110)

106 Precision grip depends almost entirely on the integrity
     of the ulnar nerve

     BECAUSE

     the ulnar nerve innervates most of the small muscles
     of the hand.

     Objectives   7; III; 3; (3.7).   7; IV; 1; (1.5).   *0.55*   91   D
                                                                  0.50   71

107 Opposition of the thumb is relatively unaffected
     after an ulnar nerve lesion

     BECAUSE

     the intrinsic muscles of the thumb are mainly supplied
     by the median nerve.

     Objectives   7; III; 3; (3.5).   7; IV; 1; (1.5).   *0.50*   54   A
                                                                  0.39   74

108 Lesions of the ulnar nerve at the wrist lead to hyper-
     extension at the metacarpophalangeal joint of the
     little finger

     BECAUSE

     the interosseous and lumbrical muscles normally flex
     at this metacarpophalangeal joint.

     Objectives   7; III; 3; (3.5).   7; IV; 1; (1.5).   *0.47*   42   A

109 An ulnar nerve lesion at the wrist affects the power grip

BECAUSE

it is no longer possible to oppose the thumb.

Objectives 7; III; 3; (3.7). 7; IV; 1; (1.5). *0.47* 85 C

110 Extension at the interphalangeal joint of the thumb is lost following section of the radial nerve

BECAUSE

the extensor pollicis brevis is supplied by the radial nerve.

Objectives 7; III; 3; (3.1). 7; IV; 1; (1.5). *0.42* 53 B

E(iv) Nerves and blood vessels (Questions 111–121)

111 Lesions of the C8 and T1 nerve roots produce a motor deficit indistinguishable from an ulnar nerve lesion

BECAUSE

the ulnar nerve is a direct branch of the lower trunk of the brachial plexus.

Objectives 7; IV; 1; (1.1), (1.5), (1.6). *0.26* 12 E

112 Lesions of the C5 and C6 nerve roots do not impair abduction at the shoulder

BECAUSE

these roots are distributed to the flexor muscles of the arm.

Objectives 7; IV; 1; (1.6). *0.30* 56 D
0.52 77

113 Division of the superficial branch of the radial nerve
     at the level of the elbow joint results in partial loss
     of supination

     BECAUSE

     the brachioradialis and supinator muscles are
     paralysed.

     Objectives   7; IV; 1.   *0.53*   73   E
                                                   0.58   73

114 Division of the radial nerve in the axilla results in loss
     of active extension at the elbow joint

     BECAUSE

     all the extensors of the elbow joint are innervated by
     the radial nerve.

     Objectives   7; IV; 1; (1.2), (1.5).   *0.51*   94   A
                                                   0.63   89

115 In a radial nerve lesion in the axilla supination is
     weakened

     BECAUSE

     the biceps muscle is paralysed.

     Objectives   7; IV; 1; (1.2), (1.5).   *0.55*   91   C

116 To test for the integrity of the radial nerve one would
     test mainly for motor loss

     BECAUSE

     sensory loss is small and unimportant due to overlap
     from adjacent nerves.

     Objectives   7; IV; 1.   *0.33*   48   A

117 Hyperextension at the metacarpophalangeal joints following an ulnar nerve lesion is more marked on the medial side of the hand

BECAUSE

the flexor digitorum superficialis tendons and the lumbricals of the ring and little fingers are supplied by the ulnar nerve.

Objectives   7; IV; 1; (1.5).   *0.40*   57   C

118 Lesions of the median nerve in the cubital fossa produce no cutaneous sensory loss in the forearm

BECAUSE

the median nerve gives off no branches until it reaches the hand.

Objectives   7; IV; 1; (1.5).   *0.36*   45   C
                                                       0.37   75
                                                       0.58   60

119 The skin on the palmar surface of the radial 3½ fingers is segmentally supplied by T1

BECAUSE

the small muscles underlying this part of the hand are supplied by the T1 segment.

Objectives   7; IV; 1; (1.6).   *0.51*   80   D

120 Compression of structures deep to the flexor retinaculum may result in paralysis of the thenar muscles

BECAUSE

the ulnar nerve is closely related to the flexor retinaculum.

Objectives   7; IV; 1; (1.2), (1.5).   *0.48*   77   B
                                                       0.35   66
                                                       0.23   53

121   The superficial and deep veins of the forearm do not
      intercommunicate

      BECAUSE

      all the blood from these veins eventually drains into
      the axillary vein.

      Objectives   7; IV; 2; (2.10).   *0.25*   71   D

Type K items (Hubbard and Clemans, 1961) (Questions 122–188)

These items are in the form of incomplete statements and call for recognition
of one or more correct responses which are grouped in various ways. The
instruction for these items is as follows:

> *For each of the incomplete statements below, one or more of the*
> *completions given is correct. Responses should be chosen as follows:*
> *(A) if only completions (i), (ii) and (iii) are correct,*
> *(B) if only completions (i) and (iii) are correct,*
> *(C) if only completions (ii) and (iv) are correct,*
> *(D) if only completion (iv) is correct,*
> *(E) if all completions are correct.*

K(i) Shoulder region (Questions 122–136)

122   The pectoralis major muscle

      (i)   is a flexor of the arm.
      (ii) is capable of extending the arm from the flexed
            position.
      (iii) medially rotates the arm.
      (iv) is an important adductor of the arm.

      Objectives   7; I; 2; (2.2).   *0.24*   34   E
                                                        0.50   83

123   In full abduction of the upper limb there must be

      (i)   lateral (upward) rotation of the scapula.
      (ii) retraction of the scapula.
      (iii) lateral rotation of the humerus.
      (iv) depression of the scapula.

      Objectives   7 I; 3; (3.6).   *0.38*   87   B

124   The axilla

(i)   has an anterior wall formed by both pectoralis
       major and minor muscles.
(ii) contains lymph nodes draining the breast and all
       the upper limb.
(iii) has an apex which reaches the external border of
       the first rib.
(iv) has a posterior wall formed by the latissimus
       dorsi and teres major muscles only.

Objectives   7; I; 2; (2.1).   7; IV; 2; (2.11).   *0.38*   59   A

125   The coraco-acromial arch

(i)   deepens the socket of the shoulder joint.
(ii) includes the coraco-clavicular ligament.
(iii) limits abduction of the arm.
(iv) may protect the shoulder joint from injury.

Objectives   7; I; 1.   *0.35*   29   D
                                                        0.35   37

126   The pectoralis major and latissimus dorsi muscles
      stand out in

(i)   pushing upwards.
(ii) shrugging the shoulder.
(iii) elevating the arm.
(iv) climbing a rope.

Objectives   7; I; 2; (2.3).   7; I; 3; (3.5).   *0.46*   77   D

127   Important protractors of the scapula include the

(i)   subscapularis.
(ii) teres major.
(iii) latissimus dorsi.
(iv) serratus anterior.

Objectives   7; I; 2; (2.3).   *0.30*   71   D

128   The coracoid process

  (i)   has its own epiphyseal centre(s).
  (ii) is palpable below the junction of the lateral and
        intermediate thirds of the clavicle.
  (iii) is attached by ligaments to both scapula and
        clavicle.
  (iv) is covered by the deltoid.

  Objectives   7; 1; 1; (1.2), (1.7).   7; 1; 3; (3.1).   *0.42*   29   E

129   Lateral rotation of the arm is produced by the

  (i)   pectoralis major.
  (ii) pectoralis minor.
  (iii) teres major.
  (iv) teres minor.

  Objectives   7; 1; 2; (2.3).   *0.68*   86   D

130   To make the latissimus dorsi stand out by contraction
       against resistance it would be reasonable to ask the
       subject to

  (i)   press forward with his hand against a flat vertical
        surface.
  (ii) lift his arm above his head.
  (iii) put his hand behind his head.
  (iv) put his hand on his hip and press downwards.

  Objectives   7; 1; 2.   7; 1; 3; (3.5).   *0.46*   65   D

131   The acromio-clavicular joint

  (i)   is strengthened by the coraco-clavicular ligament.
  (ii) is strengthened by the coraco-acromial ligament.
  (iii) is a plane synovial joint.
  (iv) involves the posterior surface of the clavicle.

  Objectives   7; 1; 1; (1.7), (1.12).   *0.51*   45   B
                                                                0.27   65

132 The clavicle

(i)   has a ligamentous attachment to its fellow of
      the opposite side.
(ii) is attached to the articular disc of the sterno-
      clavicular joint.
(iii) is attached to the first costal cartilage.
(iv) articulates with the first costal cartilage.

Objectives   7; I; 1; (1.3), (1.7).   *0.36*   43   E

133 The sterno-clavicular joint

(i)   is formed by the clavicle, sternum and first rib.
(ii) is strengthened medially by the costo-clavicular
      ligament.
(iii) is fixed during elevation of the scapula.
(iv) has an articular disc dividing its cavity into two.

Objectives   7; I; 1; (1.7), (1.10), (1.12).   *0.38*   34   D

134 The acromion

(i)   has an articular facet facing upwards and laterally.
(ii) tends to be driven over the clavicle in a fall on
      the outstretched hand.
(iii) forms the most lateral bony point of the shoulder
      region.
(iv) articulates with the clavicle at a plane joint.

Objectives   7; I; 1; (1.2), (1.3), (1.7), (1.11).
             7; I; 3; (3.1), (3.4)   *0.44*   62   D

135 If the serratus anterior muscle is paralysed, the
     patient may have difficulty in

(i)   flexing at the gleno-humeral joint.
(ii) elevating the upper limb above 90°.
(iii) bracing the shoulders.
(iv) punching someone.

Objectives   7; I; 2; (2.2), (2.3).   7; I; 3; (3.5).   *0.44*   50   C
                                                        0.48   58

136 Lateral (upward) rotation of the scapula is produced
by the

(i) serratus anterior.
(ii) levator scapulae.
(iii) trapezius.
(iv) rhomboids.

Objectives 7; I; 2; (2.3).   *0.60*   65   B

K(ii) Shoulder region including nerve supply (Questions 137–140)

137 Division of the long thoracic nerve is indicated by

(i) inability to retract the scapula.
(ii) wasting of the pectoralis major muscle.
(iii) weakness of humeral adduction.
(iv) 'winging' of the scapula.

Objectives 7; I; 2.   7; IV; 1.   *0.60*   82   D

138 Dislocation of the shoulder joint

(i) is most common inferiorly where the capsule has
no muscle attachments.
(ii) usually results from excessive lateral rotation.
(iii) more commonly damages the axillary nerve than
any other.
(iv) more commonly damages the musculo-cutaneous
nerve than any other.

Objectives 7; I; 1.   7; IV; 1; (1.2).   −   −   B

139 The pectoralis major muscle

(i) adducts the humerus.
(ii) medially rotates the humerus.
(iii) flexes at the gleno-humeral joint.
(iv) is supplied by the median nerve.

Objectives 7; I; 2; (2.2).   7; IV; 1.   −   −   A

140   In a patient with a fracture of the surgical neck of the
      humerus

      (i)   the capsule of the shoulder joint will almost
            certainly be torn.
      (ii) the nerve most likely to be damaged is the
            axillary nerve.
      (iii) the brachialis muscle will probably be torn.
      (iv) a muscle likely to be paralysed is the deltoid.

      Objectives   7; I; 1; (1.6).   7; II; 1; (1.11).
                   7; IV; 1; (1.2), (1.5).   *0.28*   94   C

K(iii) Elbow region (Questions 141–145)

141   The head of the radius

      (i)   forms a pivot joint with the capitulum.
      (ii) articulates with the humerus and the ulna at a
            common joint cavity.
      (iii) gives attachment to the anular ligament.
      (iv) is palpable just below the lateral epicondyle.

      Objectives   7; II; 1; (1.1), (1.3), (1.4), (1.10).
                   7; III; 1; (1.3), (1.4).   *0.32*   44   C

142   Extension at the elbow joint

      (i)   places the humeral epicondyles and the tip of the
            olecranon process in the position of the angles of
            an equilateral triangle.
      (ii) relaxes the ulnar and radial collateral joint
            ligaments.
      (iii) tightens the anular ligament.
      (iv) weakens the movement of supination.

      Objectives   7; II; 1; (1.1), (1.2), (1.9), (1.10), (1.12).
                   7; III; 1; (1.6).   *0.46*   65   D

143   At the elbow joint

    (i)   flexion is assisted by the pronator teres muscle.
    (ii) the radial collateral ligament is firmly attached
         to the radius.
    (iii) extension is assisted by the anconeus muscle.
    (iv) the ulnar collateral ligament has humero-ulnar
         and humero-radial bands.

Objectives   7; II; 1; (1.9), (1.11).   *0.32*   51   B

144   The elbow joint

    (i)   is a hinge joint.
    (ii) is indicated by a horizontal line joining the two
         epicondyles.
    (iii) shares a joint cavity with the proximal radio-ulnar
         joint.
    (iv) has a strong posterior capsule.

Objectives   7; II; 1; (1.4), (1.9).   *0.18*   39   B
                                            0.32   31

145   The flexors at the elbow joint include

    (i)   the brachialis, which is chiefly used.
    (ii) the biceps brachii, particularly when the forearm
         is pronated.
    (iii) the pronator teres.
    (iv) the brachioradialis, but only in mid-pronation/
         supination.

Objectives   7; II; 1; (1.13).   *0.32*   33   B

K(iv) Forearm, wrist and hand (Questions 146–158)

146   The inferior radio-ulnar joint

    (i)   communicates with the wrist joint.
    (ii) is a fibrous joint.
    (iii) is strengthened by the flexor retinaculum.

(iv) is strengthened by a fibrocartilaginous disc
running from the styloid process of the ulna
to the radius.

Objectives   7; III; 1; (1.3), (1.4).   *0.45*   61   D

147   From a fully supinated position, pronation of the
forearm can be achieved by the

(i)   pronator quadratus.
(ii) brachioradialis.
(iii) pronator teres.
(iv) biceps brachii.

Objectives   7; III; 1; (1.5).   *0.47*   26   A

148   The articulating surfaces at the wrist joint include the

(i)   distal surface of the radio-ulnar articular disc.
(ii) proximal surface of the lunate.
(iii) proximal surface of the scaphoid.
(iv) head of the ulna.

Objectives   7; III; 2; (2.2).   *0.52*   88   A
                                                               0.39   88

149   Flexion of the hand occurs at the

(i)   radiocarpal joint.
(ii) midcarpal joint only.
(iii) radiocarpal and intercarpal joints.
(iv) radiocarpal and midcarpal joints.

Objectives   7; III; 2; (2.6).   *0.36*   87   D
                                                               0.67   97
                                                               0.32   96

150   At the metacarpophalangeal joints of the fingers

(i)   extension is produced by the extensor digitorum.
(ii) flexion is produced by the lumbricals.

   (iii) flexion is produced by flexor digitorum
         profundus.
   (iv) abduction is produced by palmar interossei.

   Objectives   7; III; 3; (3.1), (3.5).   *0.76*   91   **A**

151 In relation to the flexor tendons of the index finger,

   (i)   each of the tendons of flexor digitorum profundus
         and superficialis has a separate sheath.
   (ii) a single digital synovial sheath is proximally
         continuous with a common synovial flexor
         sheath.
   (iii) the tendons of flexor digitorum profundus and
         superficialis are both flexors of the distal
         interphalangeal joint.
   (iv) the tendon of flexor digitorum superficialis is
         perforated by the tendon of flexor digitorum
         profundus.

   Objectives   7; III; 3; (3.1).   *0.50*   80   **D**
                                                         0.64   88

152 A digital synovial sheath

   (i)   surrounds only the tendon of the flexor digitorum
         profundus in the index finger.
   (ii) extends distally as far as the base of the distal
         phalanx.
   (iii) surrounds only the tendon of the flexor digitorum
         profundus in the little finger.
   (iv) in the little finger normally communicates with
         the common synovial sheath.

   Objectives   7; III; 3; (3.1).   —   —   **C**

153 To demonstrate the surface marking of the flexor
     retinaculum it would be relevant to seek the position
     of the following structures:

   (i)   proximal flexion crease of wrist.

(ii) styloid process of ulna.
(iii) styloid process of radius.
(iv) pisiform bone.

Objectives   7; III; 3; (3.3).   *0.55*   76   D

154 Opposition of the thumb involves

(i)   abduction at the carpometacarpal joint.
(ii) flexion at the carpometacarpal joint.
(iii) abduction at the metacarpophalangeal joint.
(iv) medial rotation at the carpometacarpal joint.

Objectives   7; III; 3.   *0.45*   55   E

155 The lumbricals

(i)   flex at the interphalangeal joints.
(ii) flex at the metacarpophalangeal joints.
(iii) are all supplied by the ulnar nerve.
(iv) insert into digital extensor expansions.

Objectives   7; III; 3; (3.4), (3.5).   *0.62*   82   C
                                                   0.64   90

156 At the metacarpophalangeal joint of the index finger

(i)   extension is produced by the extensor digitorum.
(ii) flexion is produced by the 1st lumbrical.
(iii) flexion is produced by flexor digitorum
      profundus.
(iv) abduction is produced by the 1st palmar interosseous.

Objectives   7; III; 3; (3.1), (3.5).   *0.48*   59   A
                                                   0.54   84
                                                   0.63   78

157 Extension of the thumb is aided by the

(i)   first lumbrical muscle.
(ii) first palmar interosseous muscle.

(iii) first dorsal interosseous muscle.
(iv) abductor pollicis longus muscle.

Objectives   7; III; 3; (3.1), (3.5).   *0.19*   83   D
                                                          0.37   79

158   The flexor synovial sheath of the 5th digit

(i)   does not extend proximal to the 5th
      metacarpophalangeal joint.
(ii) does not extend proximal to the proximal
      interphalangeal joint.
(iii) is generally absent.
(iv) extends deep to the abductor and flexor digiti
      minimi muscles.

Objectives   7; III; 3; (3.1)   *0.38*   80   D
                                                          0.44   79

K(v) Forearm, wrist and hand, including nerve supply (Questions 159–164)

159   The flexor carpi radialis muscle

(i)   flexes the hand.
(ii) abducts the hand.
(iii) is supplied by the median nerve.
(iv) has a tendon surrounded by a synovial sheath.

Objectives   7; III; 2; (2.4), (2.5), (2.6).
             7; IV; 1; (1.5).   *0.20*   59   E

160   The following structures cross superficial to the
      flexor retinaculum of the wrist:

(i)   the ulnar nerve.
(ii) the median nerve.
(iii) the tendon of the palmaris longus muscle.
(iv) the tendon of the flexor carpi radialis muscle.

Objectives   7; III; 2; (2.4), (2.7).   7; IV; 1; (1.2).   *0.57*   96   B

161   The supinator muscle

(i)   is supplied by the radial nerve.
(ii) is attached to the ulna.
(iii) is attached to the radius.
(iv) assists in flexion of the elbow.

Objectives   7; III; 1;(1.5).   7; IV; 1;(1.5).   *0.22*   85   A

162   The flexor carpi radialis muscle is

(i)   a flexor of the wrist.
(ii) an abductor of the wrist.
(iii) supplied by the median nerve.
(iv) involved in pronation.

Objectives   7; III; 2;(2.5).   7; IV; 1;(1.5).   –   –   E

163   Adduction of the hand

(i)   is limited by the radial styloid.
(ii) involves the radial nerve but not the ulnar nerve.
(iii) involves movement of the hand to the radial side.
(iv) is limited by the radial collateral ligament of the
      wrist.

Objectives   7; III; 2;(2.6).   7; IV; 1;(1.5).   *0.43*   83   D

164   The flexor carpi ulnaris muscle

(i)   adducts (ulnar deviation) the hand at the wrist
      joint.
(ii) flexes the hand.
(iii) is supplied by the ulnar nerve.
(iv) flexes the little finger.

Objectives   7; III; 2;(2.4–2.6).   *0.42*   86   A

K(vi) Nerves and blood vessels (Questions 165–188)

165   The lateral cord of the brachial plexus

   (i)   contains motor fibres only.
   (ii) gives rise to the musculo-cutaneous nerve.
   (iii) may be damaged in fractures of the surgical neck
         of the humerus.
   (iv) may contribute to the ulnar nerve.

   Objectives   7; IV; 1.   0.44   61   C
                                                       0.43   69

166   Medial rotators of the humerus are innervated through

   (i)   the dorsal scapular nerve.
   (ii) the posterior cord of the brachial plexus.
   (iii) the long thoracic nerve.
   (iv) cord segments C7 and C8.

   Objectives   7; IV; 1.   0.17   38   C

167   A wrist lesion of the ulnar nerve results in

   (i)   inability to spread the fingers.
   (ii) inability to oppose the thumb.
   (iii) loss of sweating including the palmar aspect of
         the little finger.
   (iv) inability to extend at the 5th metacarpophalangeal
         joint.

   Objectives   7; IV; 1.   0.39   82   B

168   The radial nerve

   (i)   pierces the lateral intermuscular septum.
   (ii) runs part of its course in a groove between the
         brachioradialis and the brachialis.
   (iii) divides into superficial and deep branches.
   (iv) supplies the adductor pollicis muscle.

   Objectives   7; IV; 1; (1.2), (1.5).   0.36   68   A

169   The radial nerve

  (i)   is the only branch of the posterior cord which
        reaches the forearm.
  (ii) has the root values C5 + 6.
  (iii) lies in contact with the humerus between the
        lateral and medial heads of triceps.
  (iv) has no area of cutaneous supply above the
        elbow.

Objectives   7; IV; 1; (1.2), (1.3), (1.5).   *0.52*   83   B

170   The median nerve

  (i)   receives contributions from both the lateral and
        medial cords of the brachial plexus.
  (ii) supplies a muscle which supinates the forearm.
  (iii) supplies no skin in the arm.
  (iv) lies posterior to the medial epicondyle at the
        elbow.

Objectives   7; IV; 1; (1.1), (1.2), (1.5).   *0.67*   80   B

171   Division of the ulnar nerve at the level of the medial
      epicondyle results in

  (i)   anaesthesia of the medial side of the forearm.
  (ii) anaesthesia on the medial side of the dorsum of
        the hand.
  (iii) paralysis of the extensor carpi ulnaris muscle.
  (iv) loss of abduction of the index finger.

Objectives   7; IV; 1; (1.5).   *0.45*   70   C

172   The ulnar nerve supplies the

  (i)   medial half of flexor digitorum superficialis.
  (ii) lumbrical muscle to the little finger.
  (iii) abductor pollicis brevis.
  (iv) first dorsal interosseous muscle.

Objectives   7; IV; 1; (1.5).   —   —   C

173   Muscles innervated by the median nerve include the

   (i)   palmaris brevis.
   (ii) opponens pollicis.
   (iii) adductor pollicis.
   (iv) first lumbrical.

   Objectives   7; IV; 1; (1.5).   0.55   83   C

174   The radial nerve through its branches

   (i)   is distributed to all extensors of the elbow.
   (ii) is distributed to the supinator muscle.
   (iii) is distributed to all extensors of the wrist.
   (iv) has no cutaneous sensory distribution.

   Objectives   7; IV; 1; (1.5).   0.58   76   A
                                                    0.41   92
                                                    0.57   85

175. Division of the musculocutaneous nerve may result in

   (i)   anaesthesia of the lateral side of the forearm and
         the whole of the thumb.
   (ii) weakness of supination.
   (iii) weakness of pronation.
   (iv) weakness of elbow flexion.

   Objectives   7; IV; 1; (1.5).   0.53   72   C
                                                    0.53   63

176   Damage to cord segment C6 could result in

   (i)   difficulty in pronation.
   (ii) difficulty in eliciting the biceps tendon jerk.
   (iii) difficulty in eliciting the brachioradialis tendon
         jerk.
   (iv) complete loss of sensation in the skin superficial
         to the biceps.

   Objectives   7; IV; 1; (1.6).   0.39   84   A

177 Successful testing of the brachioradialis jerk

   (i)   can be best achieved with the forearm fully
         supinated.
   (ii) indicates integrity of the radial nerve.
   (iii) involves tapping over the lateral border of the
         'anatomical snuff-box'.
   (iv) indicates integrity of the 5th and 6th cervical
         spinal cord segments.

Objectives   7; IV; 1; (1.6).   0.28   58   C
                                                 0.28   47

178 The axillary lymph nodes normally receive vessels
     from the

   (i)   face and neck.
   (ii) lateral part of the mammary gland.
   (iii) brachial lymph nodes.
   (iv) skin of the trunk above the umbilicus.

Objectives   7; IV; 2; (2.11).   0.26   61   C

179 The median cubital vein

   (i)   is an anastomotic channel.
   (ii) is crossed superficially by the bicipital
         aponeurosis.
   (iii) communicates with the deep veins of the forearm.
   (iv) runs upwards and laterally from the cephalic to
         the basilic vein.

Objectives   7; IV; 2; (2.6), (2.7), (2.10).   0.44   67   B

180 The basilic vein

   (i)   pierces the deep fascia.
   (ii) is medial to the cephalic vein.
   (iii) helps to form the axillary vein.
   (iv) lies posterior to the medial epicondyle of the
         humerus.

Objectives   7; IV; 2; (2.6).   0.38   95   A

181 Arterial pulsation in the upper limb can normally be felt

(i) directly lateral to the flexor carpi radialis tendon.
(ii) directly lateral to the flexor carpi ulnaris tendon.
(iii) over the scaphoid and trapezium.
(iv) posteriorly on the distal end of the radius.

Objectives   7; IV; 2.   *0.38*   43   A

182 The brachial artery

(i) begins at the lower border of teres major.
(ii) accompanies the ulnar nerve throughout the arm.
(iii) accompanies the median nerve throughout the arm.
(iv) usually divides just above the elbow joint.

Objectives   7; IV; 2; (2.1).   *0.44*   48   B

183 The axillary artery commences

(i) at the outer border of the first rib.
(ii) at the lower border of the pectoralis minor muscle.
(iii) in the apex of the axilla.
(iv) at the sterno-clavicular joint.

Objectives   7; IV; 2; (2.1).   *0.36*   65   B

184 The axillary artery

(i) begins at the superior border of the teres major.
(ii) has its corresponding vein on its medial side.
(iii) lies posterior to the radial nerve.
(iv) terminates at the inferior border of the teres major.

Objectives   7; IV; 2; (2.1).   *0.53*   81   C
                                                    0.39   54

185 Firm compression of the brachial artery in the mid-part
of the arm

(i) would result in very severe depletion of blood
supply of the hand.
(ii) could be produced by pressure against the
mid-shaft of the humerus.
(iii) would not necessarily deprive the upper limb of
venous return.
(iv) would have little or no effect on the radial pulse.

Objectives   7; IV; 2; (2.1), (2.5), (2.6).   *0.42*   69   A

186 The axillary lymph nodes normally receive vessels from

(i)   the whole upper limb.
(ii) the lateral part of the mammary gland.
(iii) the skin of the back of the trunk above the iliac
crest.
(iv) the skin of the front of the trunk above the
umbilicus.

Objectives   7; IV; 2; (2.11).   *0.32*   80   E
                                                   0.25   55

187 The axillary nerve

(i)   leaves the axilla at the lower border of the
subscapularis muscle.
(ii) supplies the capsule of the gleno-humeral joint.
(iii) supplies the deltoid muscle.
(iv) gives off the musculo-cutaneous nerve.

Objectives   7; IV; 1.   —   —   A

188 The skin comprising dermatome C5

(i)   obtains its sensory supply only from C5 spinal
nerve.
(ii) obtains its sensory supply only via the dorsal
roots of spinal nerves.
(iii) receives C5 fibres only via the axillary nerve.
(iv) receives its sensory supply only via the upper
trunk of the brachial plexus.

Objectives   7; IV; 1.   *0.30*   16   C

Chapter 3

Head and Neck

Type A items (Hubbard and Clemans, 1961) (Questions 189–258)

These involve choosing one correct answer from five available choices. The instruction for these items is as follows:

Each of the incomplete statements below is followed by five suggested answers or completions. Select the one which is best in each case.

A(i) Cranio-cerebral topography, eye, ear and nose (Questions 189–211)

189 In relation to the skull the

(A) bones contain only fatty marrow.
(B) frontal and parietal bones articulate at the lambdoid suture.
(C) bones are all paired.
(D) hypophysial fossa is not visible on a lateral radiograph.
(E) posterior cranial fossa is floored by the occipital and temporal bones.

Objectives   8; I; 1; (1.1), (1.2), (1.7).   *0.34*   81   E
                                                    0.28   88
                                                    0.44   84

190 The foramen magnum transmits

(A) the vertebral arteries.
(B) the hypoglossal nerves.
(C) the abducent nerves.
(D) the pons.
(E) none of the above.

Objectives   8; I; 1; (1.1–1.6).   8; I; 3; (3.1).   *0.44*   92   A

191 At the lower end of the pons, the following nerves are attached:

(A) trigeminal.
(B) glossopharyngeal.
(C) hypoglossal.
(D) abducent.
(E) none of the above.

Objectives   8; I; 1; (1.3), (1.4).   8; VIII; 1; (1.1).   –   –   D

192 The foramen ovale is situated in

(A) the frontal bone.
(B) the temporal bone.
(C) the ethmoid bone.
(D) the maxillary bone.
(E) none of the above.

Objectives   8; I; 1; (1.1), (1.5).   *0.31*   45   E
                                                           0.25   43
                                                           0.31   65
                                                           0.40   79

193 The foramen ovale transmits

(A) the maxillary nerve.
(B) the maxillary artery.
(C) the middle meningeal artery.
(D) the mandibular nerve.
(E) both the maxillary nerve and the middle
    meningeal artery.

Objectives   8; I; 1; (1.1), (1.5), (1.6).   *0.51*   97   D

194   The greater wing of the sphenoid

    (A) forms part of the floor of the anterior cranial
        fossa.
    (B) is the superior boundary of the superior orbital
        fissure.
    (C) contains the foramina of exit of both maxillary
        and mandibular nerves.
    (D) is in contact with the frontal lobe of the cerebrum.
    (E) forms part of the lateral wall of the nose.

Objectives   8; I; 1; (1.1–1.6).   8; III; 1; (1.1).   *0.46*   76   C

195   The mastoid process is part of

    (A) the occipital bone.
    (B) the sphenoid bone.
    (C) the zygomatic bone.
    (D) the parietal bone.
    (E) none of the above bones.

Objectives   8; I; 1; (1.1), (1.7).   –   –   E

196   The first teeth of the permanent dentition to erupt are

    (A) the central incisors.
    (B) the first molars.
    (C) the first premolars.
    (D) the canines.
    (E) none of the above.

Objectives   8; IX; 1; (1.1).   –   –   B

197   The lateral margin of the orbital aperture is formed
    by the

    (A) zygomatic bone.
    (B) frontal bone.
    (C) zygomatic and frontal bones.
    (D) zygomatic and sphenoid bones.
    (E) frontal and sphenoid bones.

Objectives   8; I; 1; (1.5), (1.6), (1.8).   –   –   C

198   The superior orbital fissure is bounded by

    (A) the maxilla and greater wing of the sphenoid.
    (B) the maxilla and lesser wing of the sphenoid.
    (C) the lesser wing of the sphenoid and the ethmoid.
    (D) the lesser and greater wings of the sphenoid.
    (E) none of the above bones.

Objectives   8; I; 1; (1.1), (1.5), (1.8).   *0.50* 80   D

199   The superior orbital fissure transmits

    (A) the optic nerve.
    (B) the ophthalmic artery.
    (C) the trochlear nerve.
    (D) the maxillary nerve.
    (E) none of the above.

Objectives   8; I; 1; (1.5), (1.6).   8; I; 3; (3.5).
            8; II; 1; (1.1).   –   –   C

200   The superior oblique muscle

    (A) is supplied by the abducent nerve.
    (B) is supplied by the oculomotor nerve.
    (C) produces upwards eye movement.
    (D) is supplied by the trochlear nerve.
    (E) elevates the upper eyelid.

Objectives   8; VIII; 1; (1.5), (1.9).   *0.38* 95   D

201   The superior oblique muscle

    (A) elevates the eye.
    (B) produces downward and outward movement of
        the eye.
    (C) adducts the eye.
    (D) is supplied by the ophthalmic division of the
        trigeminal nerve.
    (E) is supplied by the oculomotor nerve.

Objectives   8; VIII; 1; (1.5), (1.9).   *0.34* 90   B

202   The size of the pupil

(A) is independent of parasympathetic activity.
(B) is independent of sympathetic activity.
(C) depends on the state of contraction of the ciliary
      muscle.
(D) alters with light and accommodation stimuli.
(E) alters only with light stimuli.

Objectives   8; I; 4; (4.2—4.6).
              8; II; 1; (1.1), (1.5), (1.10).
              8; VIII; 1; (1.4).   *0.28*   81   D
                                                      0.33   65

203. The optic disc

(A) lies at the junction of the sclera and the cornea.
(B) lies at the posterior pole of the eyeball.
(C) contains the central artery of the retina.
(D) is particularly sensitive to light.
(E) contains retinal receptors.

Objectives   8; II; 1; (1.1—1.6).   8; II; (1.9), (1.10).   *0.41*   71   C

204   The optic nerve

(A) enters the posterior pole of the eyeball.
(B) is nourished by the aqueous humour.
(C) has a covering of three meningeal layers.
(D) enters the orbit through the superior orbital
      fissure.
(E) has none of the above properties.

Objectives   8; I; 1.   8; II; 1.   *0.31*   63   C

205   The naso-lacrimal duct opens into

(A) the hiatus semilunaris.
(B) the middle meatus.
(C) the infundibulum.

(D) the atrium.
(E) none of the above.

Objectives 8; I; 1; (1.1), (1.7).
8; III; 1; (1.1), (1.2), (1.4).   —   —   E

206 The naso-lacrimal duct opens into the

(A) ethmoidal sinuses.
(B) maxillary sinus.
(C) superior meatus.
(D) inferior meatus.
(E) middle meatus.

Objectives 8; II; 1; (1.1), (1.7).
8; III; 1; (1.1), (1.2), (1.4).   *0.62*   84   D

207 The middle ear

(A) communicates with the oropharynx via the
auditory tube.
(B) communicates with ethmoid air cells.
(C) has a round window closed by the base of the
stapes.
(D) is traversed by the vestibulocochlear nerve.
(E) has none of the above properties.

Objectives 8; III; 1; (1.2), (1.3).   8; V; 1; (1.1).
8; VII; 1; (1.2), (1.3), (1.8).   *0.59*   69   E

208 The middle nasal concha is part of

(A) the vomer.
(B) the maxilla.
(C) the ethmoid.
(D) the sphenoid.
(E) none of the above.

Objectives 8; I; 1; (1.1), (1.8).
8; III; 1; (1,1), (1.3).   —   —   C

209   The posterior border of the septal cartilage of the
nose is connected with the

(A) middle concha.
(B) body of the sphenoid.
(C) perpendicular plate of the ethmoid and vomer.
(D) vomer only.
(E) perpendicular plate of the ethmoid only.

Objectives   8; I; 1; (1.1), (1.8).   8; III; 1; (1.1).   *0.39*   81   C

210   The paranasal sinuses

(A) drain into the middle and inferior meatuses.
(B) are well developed at birth.
(C) may become infected from the nasal cavity.
(D) depend mainly on gravity for drainage.
(E) are lined with cuboidal epithelium.

Objectives   8; III; 1; (1.1–1.4).   8; IX; 1; (1.1).   *0.44*   88   C

211   The paranasal sinuses not drained by gravity in the
erect posture are the

(A) frontal.
(B) sphenoidal.
(C) maxillary.
(D) anterior cells of ethmoid.
(E) posterior cells of ethmoid.

Objectives   8; I; 1; (1.1), (1.8).   8; II; 1; (1.1–1.5).   –   –   C

A(ii) Musculo-skeletal structures in the neck (Questions 212–221)

212   The sternocleidomastoid muscle

(A) is in the floor of the posterior triangle.
(B) receives its main motor innervation from the
cervical plexus.

(C) forms the anterior border of the posterior triangle.
(D) rotates the face to its own side.
(E) has none of the above properties.

Objectives   8; I; 2; (2.2), (2.5), (2.6).   —   —,   C

213   The right sternocleidomastoid muscle acting alone

(A) fully extends the cervical vertebrae.
(B) fully flexes the cervical vertebrae.
(C) turns the head upwards and to the right.
(D) turns the head upward and to the left.
(E) performs all the above actions.

Objectives   8; I; 2; (2.1–2.7).   0.28   78   D

214   The sternocleidomastoid muscle is innervated by

(A) the glossopharyngeal nerve.
(B) the external branch of the accessory nerve.
(C) the superior laryngeal nerve.
(D) the vagus nerve.
(E) none of the above.

Objectives   8; I; 2; (2.2), (2.3).   8; VIII; 1; (1.5).   0.20   98   B
                                                              0.70   94

215   The trapezius muscle derives its nerve supply from the

(A) accessory nerve only.
(B) accessory plus twigs from C4 and C5.
(C) accessory plus the suprascapular nerve.
(D) accessory plus contributions from C3 to C4 through the cervical plexus.
(E) accessory plus muscular twigs from the dorsal rami of C3 and C4.

Objectives   8; I; 2.   —   —   D

216   The scalenus anterior muscle

(A) lies posterior to the subclavian artery.
(B) is immediately anterior to the vagus nerve.
(C) is immediately anterior to the phrenic nerve.
(D) lies posterior to the brachial plexus.
(E) has none of the above properties.

Objectives   8; 1; 2; (2.2), (2.3).   8; 1; 3; (3.1), (3.8).   –   –   E

217   The post-vertebral muscles

(A) have the pharynx as an immediate anterior
    relation.
(B) cause flexion of the head on the neck.
(C) cause flexion of the neck on the trunk.
(D) are the superior continuation of the erector
    spinae group.
(E) are innervated by the accessory nerve.

Objectives   8; 1; 2; (2.2), (2.4), (2.7).   –   –   D

218   The axis vertebra

(A) has only one pair of articular surfaces.
(B) has no foramen transversarium.
(C) is a typical cervical vertebra.
(D) has an odontoid process (dens) arising from its
    body.
(E) articulates with the skull by a synovial joint.

Objectives   8; 1; 2; (2.1), (2.7).   –   –   D

219   A typical cervical vertebra has

(A) a non-bifid spine.
(B) a large heart-shaped body.
(C) superior articular surfaces facing both superiorly
    and posteriorly.
(D) a foramen in the pedicle to transmit the vertebral
    artery.
(E) none of the above properties.

Objectives   8; 1; 2; (2.1).   8; 1; 3; (3.1).   *0.36*   48   C

220 The joint between the dens of the axis vertebra and the anterior arch of the atlas vertebra is a

(A) plane synovial joint.
(B) condyloid synovial joint.
(C) secondary cartilaginous joint.
(D) pivot synovial joint.
(E) fibrous joint.

Objectives 8; I; 2; (2.1), (2.7). *0.53* 69 D

221 The transverse process of the atlas vertebra

(A) can be palpated just posterior to the mastoid process.
(B) can be palpated about 2 fingers' breadth below the mastoid process.
(C) can be palpated deeply at the back of the neck.
(D) can be palpated just below and in front of the mastoid process.
(E) cannot be palpated anywhere.

Objectives 8; I; 1; (1.7). 8; I; 2; (2.1). *0.42* 33 D

A(iii) Blood vessels and lymph nodes (Questions 222–232)

222 The external carotid artery

(A) extends between the level of the upper border of the thyroid cartilage and the neck of the mandible.
(B) passes upwards superficial to the posterior belly of the digastric.
(C) lies within the substance of the submandibular salivary gland.
(D) gives off the inferior thyroid artery.
(E) has none of the above properties.

Objectives 8; I; 3; (3.1), (3.2), (3.6). 8; IV; 2; (2.1).
8; IV; 3; (3.7). *0.21* 54 A

223   The right external carotid artery

(A) arises from the aorta.
(B) arises from the subclavian artery.
(C) arises from the brachiocephalic trunk.
(D) gives off the inferior thyroid artery.
(E) has none of the above properties.

Objectives   8; I; 3; (3.1), (3.2), (3.6).   *0.25*   70   E

224   The left common carotid artery

(A) arises from the brachiocephalic trunk behind the
    sterno-clavicular joint.
(B) ascends to enter the neck lateral to the cervical
    transverse processes.
(C) is lateral to the internal jugular vein.
(D) is covered by the sternocleidomastoid.
(E) has none of the above properties.

Objectives   8; I; 2; (2.2).   8; I; 3; (3.1), (3.3), (3.10)   –   –   D

225   The internal carotid artery

(A) vertically traverses the foramen lacerum.
(B) gives off the central retinal artery.
(C) lies superficial to the sternocleidomastoid.
(D) gives off middle and anterior cerebral arteries.
(E) gives off the maxillary artery.

Objectives   8; I; 1; (1.5).   *0.33*   90   D
             8; I; 3; (3.1), (3.2), (3.5), (3.6).   *0.53*   94
             8; II; 1; (1.6).   *0.66*   86

226   The internal carotid artery

(A) arises from the brachiocephalic trunk at the
    upper border of the thyroid cartilage.
(B) ascends vertically in the neck but bends sharply
    before reaching the skull.

(C) is overlapped by the sternocleidomastoid muscle.
(D) lies lateral to the external carotid superiorly.
(E) bears no relationship to the pharynx.

Objectives 8; I; 2; (2.2).   8; I; 3; (3.1), (3.3).
8; V; 1; (1.1).   8; VI; 1; (1.8).   —   —   C

227 The vertebral artery

(A) traverses the foramina transversaria of all the
cervical vertebrae.
(B) arises from the common carotid artery.
(C) is a component of the circulus arteriosus.
(D) enters the skull through the foramen magnum.
(E) gives origin to the posterior cerebral artery.

Objectives 8; I; (1.5).   8; I; 2; (2.1).
8; I; 3; (3.1), (3.3), (3.5), (3.6).   —   —   D

228 The vertebral artery

(A) is distributed to the temporal lobe of the brain.
(B) arises from the axillary artery.
(C) turns laterally above the atlas to enter the skull.
(D) enters the vertebral canal at the level of C6.
(E) has none of the above properties.

Objectives 8; I; 1; (1.3), (1.5).   8; I; 2; (2.1).   —   —   E

229 The internal jugular vein

(A) begins in the jugular foramen as a continuation
of the straight sinus.
(B) is mainly medial to the internal carotid artery.
(C) joins the brachiocephalic vein to become the
superior vena cava.
(D) is enclosed in the carotid sheath with the vagus
nerve and the internal carotid artery.
(E) has none of the above properties.

Objectives 8; I; 1; (1.5).   8; I; 3; (3.10), (3.11).
8; VIII; 1; (1.1).   —   —   D

230   The surface marking of the internal jugular vein

   (A) is on a line joining the mastoid process to the
        sternoclavicular joint.
   (B) follows the line of the anterior border of the
        sternocleidomastoid muscle.
   (C) follows the line of the posterior border of the
        sternocleidomastoid muscle.
   (D) is on a line from the mid-point between the
        symphysis menti and the angle of the mandible
        to the mid-point of the clavicle.
   (E) is on a line from the mid-point between the
        mastoid process and the angle of the mandible
        to the sternal end of the clavicle.

   Objectives   8; l; 1; (1.7), (1.8).   8; l; 2; (2.2).   *0.31*   54   E

231   The internal jugular vein

   (A) traverses the foramen lacerum.
   (B) drains directly into the superior vena cava.
   (C) is related to deep cervical lymph nodes.
   (D) usually lies superficial to the sternocleidomastoid
        muscle.
   (E) traverses the foramen ovale.

   Objectives   8; l; 1; (1.5).   8; l; 2; (2.2).   *0.56*   68   C
                8; l; 3; (3.10), (3.11), (3.13).   *0.86*   95
                                                              0.57   91

232   Lymph nodes in the head and neck

   (A) surround muscle attachments.
   (B) drain directly into the internal jugular vein.
   (C) are distributed intracranially and extracranially.
   (D) have a parasympathetic secretomotor supply.
   (E) are distributed extracranially.

   Objectives   8; l; 3; (3.10), (3.13).   *0.47*   75   E
                                                              0.43   90
                                                              0.67   79

A(iv) Masticatory process, swallowing and phonation (Questions 233–247)

233   Attached to the articular disc of the
temporomandibular joint is the

(A) medial pterygoid muscle.
(B) temporalis muscle.
(C) lateral pterygoid muscle.
(D) masseter muscle.
(E) posterior belly of digastric.

Objectives   8; I; 2; (2.2).   8; IV; 1; (1.1), (1.2).
              8; IV; 2; (2.1), (2.5).   0.66   97   C

234   The intrinsic muscles of the tongue are supplied by the

(A) mandibular nerve.
(B) glossopharyngeal nerve.
(C) hypoglossal nerve.
(D) facial nerve.
(E) internal branch of the accessory nerve.

Objectives   8; IV; 3; (3.1), (3.2), (3.5).
              8; VIII; 1; (1.5).   0.53   91   C

235   Taste nerves from the tongue

(A) are found in facial and glossopharyngeal nerves.
(B) are found only in the facial nerve.
(C) are found only in the hypoglossal nerve.
(D) originate only in the anterior two-thirds of the
    tongue.
(E) are found only in the glossopharyngeal nerve.

Objectives   8; IV; 3; (3.1), (3.4).   8; VIII; 1; (1.2).   0.48   91   A
                                                            0.73   89
                                                            0.87   98

236   The submandibular duct opens at the region of

    (A) the second lower molar tooth.
    (B) the second upper molar tooth.
    (C) the base of the tonsil.
    (D) the base of frenulum linguae.
    (E) none of the above.

Objectives   8; IV; 3; (3.1), (3.7), (3.10).   *0.33*   77   D
                                                         0.37   80

237   The parotid gland

    (A) is supplied by the facial nerve.
    (B) has a mucous secretion.
    (C) drains into the vestibule of the oral cavity.
    (D) is supplied by the maxillary nerve.
    (E) has none of the above properties.

Objectives   8; IV; 3; (3.1), (3.7), (3.9).
               8; VIII; 1; (1.2), (1.4).   *0.60*   86   C

238   The parotid gland receives its parasympathetic
     fibres from the

    (A) vagus nerve.
    (B) lacrimal nerve.
    (C) facial nerve.
    (D) trigeminal nerve.
    (E) glossopharyngeal nerve.

Objectives   8; IV; 3; (3.7), (3.9).
               8; VIII; 1; (1.2), (1.4).   *0.61*   93   E

239   The pharynx is closely related to the

    (A) ethmoidal air sinus.
    (B) sternohyoid muscle.
    (C) sublingual gland.
    (D) prevertebral fascia.
    (E) accessory nerve.

Objectives   8; I; 1; (1.1).   8; I; 2; (2.3).
               8; IV; 3; (3.7).   8; V; 1; (1.1).   *0.36*   88   D

240 Pharyngeal constrictor muscles

(A) are composed largely of smooth muscle.
(B) all gain attachment to the hyoid bone.
(C) are innervated by the cervical plexus.
(D) gain attachment to the pharyngeal raphe.
(E) are innervated by the glossopharyngeal nerve.

Objectives 8; I; 2; (2.3).   *0.51* 61   D
            8; V; 1; (1.1), (1.2), (1.3).   *0.66* 89
            8; VIII; 1; (1.2).

241 The laryngeal part of the pharynx extends

(A) from the level of the hyoid bone to the level of
    the cricoid cartilage.
(B) from the level of the aryepiglottic fold to the
    level of the cricoid cartilage.
(C) from the level of the tip of the epiglottis to the
    level of the cricoid cartilage.
(D) opposite the bodies of cervical vertebrae 4—6.
(E) between none of the above levels.

Objectives   8; V; 1; (1.1).   8; VI; 1; (1.1), 1.3).   *0.20* 66   C
                                                        0.67 64

242 The larynx

(A) is innervated solely by the recurrent laryngeal
    nerve.
(B) has little or no lymphatic drainage.
(C) has an important function in guarding entry into
    the air passages.
(D) has a fixed relationship to the vertebral column.
(E) has a uniform type of mucosa.

Objectives   8; VI; 1; (1.1), (1.3).   —   —   C

243 The ventricle (sinus) of the larynx lies

(A) above the vestibular folds.

(B) above the vocal folds.
(C) below the vocal folds.
(D) in the vestibule.
(E) lateral to the laryngeal inlet.

Objectives   8; VI; 1; (1.3), (1.4).   *0.39*   88   B

244 The vocal folds are shortened and relaxed by

(A) the lateral cricoarytenoid muscle.
(B) the cricothyroid muscle.
(C) the posterior cricoarytenoid muscle.
(D) the thyroarytenoid muscle.
(E) none of the above.

Objectives   8; VI; 1.   *0.36*   31   D

245 Abduction of the vocal folds results from contraction
of the

(A) cricothyroid muscles.
(B) posterior cricoarytenoid muscles.
(C) vocalis muscles.
(D) thyroepiglottic muscles.
(E) lateral cricoarytenoid and transverse
     arytenoid muscles.

Objectives   8; VI; 1.   *0.46*   56   B

246 A muscle which elevates and retracts the mandible is
the

(A) temporalis.
(B) digastric.
(C) medial pterygoid.
(D) masseter.
(E) lateral pterygoid.

Objectives   8; IV; 1; (1.4).
             8; VI; 2; (2.1), (2.4), (2.5).   −   −   A

247   The temporalis muscle

(A) is supplied by the facial nerve.
(B) elevates and protracts the mandible.
(C) cannot be palpated.
(D) elevates and retracts the mandible.
(E) has none of the above properties.

Objectives   8; IV; 2; (2.1–2.5).   0.57   81   D
                                                      0.61   83

A(v) Cranial nerves (Questions 248–258)

248   The olfactory nerves

(A) contain some taste fibres.
(B) originate in the middle nasal meatus.
(C) traverse the ethmoid bone.
(D) have a secretomotor function.
(E) travel with branches of the maxillary nerve.

Objectives   8; I; 1; (1.1).   0.63   90   C
             8; III; 1; (1.1), (1.2), (1.6).   0.88   96
             8; VIII; 1; (1.2), (1.4).

249   The maxillary nerve

(A) traverses the superior orbital fissure.
(B) supplies the temporalis muscle.
(C) has cells of origin in the trigeminal ganglion.
(D) traverses the foramen ovale.
(E) innervates the facial muscles.

Objectives   8; I; 1; (1.5), (1.6), (1.10–1.12).   0.35   88   C
             8; VIII; 1; (1.2)   0.41   91

250   The main sensory innervation of the lower lip is
      derived from

(A) the hypoglossal nerve.
(B) the facial nerve.

(C) the mandibular nerve.
(D) the maxillary nerve.
(E) none of these.

Objectives   8; I; 1; (1.12).   8; VIII; 1; (1.2).   *0.39*   97   C

251   The facial nerve

(A) arises from the midbrain.
(B) leaves the cranial cavity through the foramen
     ovale.
(C) is distributed to the muscles of mastication.
(D) supplies the skin of the face.
(E) provides secretomotor innervation for the
     submandibular gland.

Objectives   8; I; 1; (1.5), (1.10–1.12).
              8; IV; 3; (3.7), (3.9).
              8; VIII; 1; (1.2), (1.4)   *0.55*   96   E

252   The facial nerve

(A) emerges from the foramen spinosum.
(B) supplies the muscles of mastication.
(C) is distributed to the skin of the face.
(D) supplies the buccinator muscle.
(E) has none of the above properties.

Objectives   8; I; 1; (1.5), (1.10), (1.12).   *0.45*   87   D

253   The vestibulocochlear nerve leaves the cranial
      cavity by the

(A) foramen rotundum.
(B) stylomastoid foramen.
(C) jugular foramen.
(D) external acoustic meatus.
(E) none of the above.

Objectives   8; I; 1; (1.4), (1.5).   *0.31*   87   E
              8; VII; 1; (1.1), (1.8), (1.9).   *0.79*   91

254 The glossopharyngeal nerve supplies

(A) the constrictor muscles.
(B) the palatopharyngeus muscle.
(C) the salpingopharyngeus muscle.
(D) the stylopharyngeus muscle.
(E) none of the above.

Objectives 8; V; 1; (1.1), (1.3). *0.60* 82 D

255 The effects of division of the left vagus nerve in the
lower third of the neck include

(A) loss of sensation in the laryngeal vestibule.
(B) inability to abduct the vocal fold.
(C) inability to adduct the vocal fold.
(D) loss of sensation in the parietal serous pericardium.
(E) none of the above.

Objectives 8; VI; 1; (1.3), (1.4), (1.6).
8; VIII; 1; (1.2), (1.7). – – B

256 The external branch of the accessory nerve

(A) enters the skull through the jugular foramen.
(B) emerges from the spinal cord with the ventral
roots of the upper cervical spinal nerves.
(C) is joined below the jugular foramen by the internal
branch of the accessory nerve.
(D) is distributed to only the sternocleidomastoid
and trapezius muscles.
(E) arises by rootlets from the side of the medulla
oblongata.

Objectives 8; I; 1; (1.4), (1.5). 8; I; 2; (2.2), (2.3).
8; VIII; 1; (1.1), (1.5). *0.33* 45 D

257 Division of the hypoglossal nerve would

(A) impair salivation.
(B) produce sensory loss over the corresponding
half of the tongue.
(C) result in impairment of taste.
(D) paralyse the tongue on the same side.

(E) have none of these effects.

Objectives   8; IV; 3; (3.4), (3.5), (3.7), (3.9), (3.19).
8; VIII; 1; (1.2), (1.4), (1.5), (1.6).   *0.51*   93   D

258   As a result of cutting one hypoglossal nerve

(A) the tongue drops down in the mouth because of
the loss of function of the mylohyoid.
(B) the tongue is retracted because of the unopposed
action of the styloglossus.
(C) the tongue, when protruded, deviates to the same
side.
(D) taste is lost on the anterior two-thirds of the
tongue.
(E) atrophy of the opposite half of the tongue occurs.

Objectives   8; IV; 2; (2.1), (2.2), (2.4).
8; IV; 3; (3.1), (3.2), (3.5), (3.19).   *0.38*   90   C

Type E items (Hubbard and Clemans, 1961) (Questions 259–289)

These items are of the assertion–reason type and there are five possible
responses. The instruction for these items is as follows:
   *Each question consists of an assertion and a reason. Responses should
   be chosen as follows:*
      *(A) if the assertion and reason are true statements and the reason is a
      correct explanation of the assertion;*
      *(B) if the assertion and reason are true statements but the reason is
      NOT a correct explanation of the assertion;*
      (C) if the assertion is true but the reason is a false statement;
      (D) if the assertion is false but the reason is a true statement;
      (E) if both assertion and reason are false statements.

E(i) Cranio-cerebral topography, eye, ear and nose (Questions 259–267)

259   The fetal skull can undergo changes of shape during
birth

BECAUSE

the fetal skull vault bones are largely cartilaginous at
birth.

Objectives   8; IX; 1; (1.1).   *0.34*   46   C

260 The constituent layers of the scalp move as one unit

BECAUSE

they are all closely adherent.

Objectives   8; I; 1; (1.10).   –   –   E

261 A lesion of the oculomotor nerve (III) results in a
medial squint

BECAUSE

the lateral rectus is paralysed and therefore
unopposed.

Objectives   8; I; 1; (1.13), (1.14).   –   –   E

262 Blocking of the central retinal artery causes blindness

BECAUSE

it is the only artery supplying the eye.

Objectives   8; I; 1; (1.6).   0.30   64   C
                                                                0.35   87

263 The superior oblique muscle depresses the eye,
rotates it laterally and intorts it

BECAUSE

its line of pull runs straight from the apex of the
orbit to the sclera in front of the equator of the eye.

Objectives   8; II; 1; (1.1), (1.13).   0.41   72   C

264 Preganglionic sympathetic fibres do not synapse in
the cervico-thoracic ganglion

BECAUSE

preganglionic sympathetic outflow is confined to
the thoracic region.

Objectives   8; I; 4; (4.2), (4.3), (4.6).   0.44   52   E
                                                                0.61   57

265   After cervico-thoracic ganglionectomy the light
reflex is lost

BECAUSE

the sphincter pupillae is innervated by sympathetic
nerve fibres.

Objectives   8; I; 1; (1.1), (1.5), (1.10).
8; I; 4; (4.2), (4.6).   *0.38*   61   E

266   The middle ear cannot communicate with the air

BECAUSE

the tympanic membrane prevents communication
between external acoustic meatus and middle ear.

Objectives   8; VII; 1; (1.1), (1.2).   *0.44*   64   D
0.61   91
0.59   79

267   Inflammation of the maxillary sinus (sinusitis) may
be associated with drainage problems

BECAUSE

the opening of the maxillary sinus is located at a
higher level than its floor.

Objectives   8; III; 1; (1.4).   −   −   A

E(ii) Musculo-skeletal structures in the neck (Questions 268–269)

268   Only flexion and extension can take place between
the atlas and axis vertebrae

BECAUSE

the alar ligaments prevent rotation between these
bones.

Objectives   8; I; 2.   *0.47*   77   F

269 The thyroid gland moves with the larynx and trachea
during swallowing

BECAUSE

it is bound to these structures by the pretracheal
fascia.

Objectives   8; V; 1; (1.4).   8; VI; 1; (1.8).   —   —   A

E(iii) Blood vessels and lymph nodes (Questions 270–271)

270 The flow of blood through the common carotid
artery may be controlled by pressing against the
carotid tubercle of C3 vertebra

BECAUSE

pulsations of this artery can be felt at the upper
level of the larynx.

Objectives   8; I; 3; (3.1), (3.3), (3.4), (3.8).   —   —   E

271 In the case of a right carotid angiogram, the left
anterior cerebral artery may be seen

BECAUSE

the anterior communicating artery may allow enough
contrast medium to pass to the opposite side.

Objectives   8; I; 3; (3.3), (3.5), (3.7), (3.9).   —   —   A

E(iv) Masticatory process, swallowing and phonation (Questions 272–282)

272 The lateral pterygoid muscle is an important
protractor of the mandible

BECAUSE

many of its fibres run backwards to insert into the
mandible and the temporomandibular articular disc.

Objectives   8; I; 1; (1.8), (1.9).
             8; IV; 2; (2.1), (2.4), (2.5).   *0.51*   77   A

273   The temporomandibular joint belongs to the class of
      cartilaginous joints

      BECAUSE

      an articular disc of fibrocartilage intervenes between
      the articular surfaces of the mandible and the
      temporal bone.

      Objectives   8; IV; 1.   —   —   D

274   Lesions of the mandibular nerve cause impairment
      of mastication

      BECAUSE

      the mandibular nerve supplies the buccinator-
      orbicularis oris muscle.

      Objectives   8; I; 1; (1.10–1.12), (1.14).   0.55   87   C
                    8; IV; 2; (2.1), (2.2), (2.5).   0.58   77
                    8; VIII; 1; (1.8).   0.75   94

275   Division of the facial nerve at the stylomastoid
      foramen does not impair mastication

      BECAUSE

      the masticatory muscles are supplied by the
      mandibular nerve.

      Objectives   8; I; 1; (1.10), (1.11), (1.13), (1.14).
                    8; IV; 2; (2.1–2.3).   8; IV; 3; (3.20).   —   —   D

276   When the motor innervation of the tongue is damaged
      unilaterally, the tongue deviates to the non-paralysed
      side on protraction

      BECAUSE

      all the tongue muscles except the palatoglossus are
      supplied by the glossopharyngeal nerve.

      Objectives   8; IV; 3; (3.1), (3.2), (3.5), (3.6), (3.19).
                    8; VIII; 1; 15, (1.6).   0.43   72   E

277 All the pharyngeal muscles are constrictor in action

BECAUSE

such action is the sole factor in the passage of food
and fluid into the oesophagus.

Objectives   8; IV; 3; (3.16).   8; V; 1; (1.1), (1.4).   *0.43*   69   E
                                                        0.41   66

278 Initiation of swallowing is largely autonomically
controlled

BECAUSE

pharyngeal musculature is mainly supplied by the
cranial accessory and vagus nerves.

Objectives   8; I; 4; (4.3).   8; IV; 3; (3.16).   *0.45*   42   D
             8; V; 1; (1.3), (1.4).   *0.46*   56
             8; VIII; 1; (1.4), (1.7).   *0.59*   65
                                                  0.48   77

279 Division of the hypoglossal nerve impairs the process
of swallowing

BECAUSE

the hypoglossal nerve supplies the muscles of the soft
palate.

Objectives   8; IV; 3; (3.4–3.6).   8; V; 1; (1.4).   *0.59*   73   C
             8; VIII; 1; (1.2), (1.6).   *0.68*   89
                                                     0.49   92

280 The vocal folds are totally paralysed by bilateral
division of the vagi at the base of the skull

BECAUSE

vagal branches supply all the intrinsic muscles of
the larynx.

Objectives   8; I; 1; (1.4).   8; VI; 1; (1.6).
             8; VIII; 1; (1.7).   *0.40*   66   A

281 The cricothyroid muscle is supplied by the
     external laryngeal nerve

     BECAUSE

     this nerve supplies all the intrinsic muscles of
     the larynx.

     Objectives   8; VI; 1; (1.6).   8; VIII; 1; (1.5).   –   –   C

282 Bilateral division of the recurrent laryngeal nerves
     usually results in difficulty in breathing

     BECAUSE

     all the intrinsic laryngeal muscles are effectively
     paralysed with the exception of one on each side
     (cricothyroid) responsible for producing increased
     tension in the vocal folds.

     Objectives   8; VI; 1; (1.5), (1.6).
                  8; VIII; 1; (1.5), (1.7)   0.37   60   A

E(v) Cranial nerves (Questions 283–289)

283 Division of the facial nerve would impair drainage of
     tear fluid

     BECAUSE

     the orbicularis oculi is paralysed.

     Objectives   8; I; 1; (1.10), (1.13), (1.14).
                  8; II; 1; (1.7), (1.15).   8; IV; 3; (3.20).
                  8; VIII; 1; (1.5), (1.6).   0.16   67   A

284 Tear secretion depends on facial nerve integrity

     BECAUSE

     the lacrimal gland gets its secretomotor innervation
     from the facial nerve.

     Objectives   8; I; 1; (1.10), (1.13), (1.14).
                  8; II; 1; (1.5), (1.7).   0.49   81   A
                  8; VIII; 1; (1.4–1.6)   0.43   86

285 Lesions of the oculomotor nerve result in impairment
of tear drainage

BECAUSE

it supplies the orbicularis oculi muscle.

Objectives   8; I; 1; (1.10), (1.11), (1.14).
             8; II; 1; (1.7), (1.13), (1.15).
             8; VIII; 1; (1.6).   —   —   E

286 Destruction of the intra-petrous part of the facial
nerve (VII) will cause an increase in the flow of tears

BECAUSE

the secretion of the lacrimal gland is controlled by
parasympathetic nerves.

Objectives   8; I; 4; (4.2).   8; II; 1; (1.7), (1.15).
             8; IV; 3; (3.20).   8; VII; 1; (1.9).
             8; VIII; 1; (1.4), (1.6).   0.54   86   D

287 If both vagus nerves were severed at the level of the
angle of the mandible there would be a lack of
secretion from the submandibular glands

BECAUSE

parasympathetic (secretomotor) impulses could not
reach the glands.

Objectives   8; I; 4; (4.2–4.4).   8; IV; 3; (3.7), (3.9).
             8; VIII; 1; (1.4), (1.6), (1.7).   0.63   86   E

288 Parasympathetic overactivity may produce a dry
mouth

BECAUSE

parasympathetic fibres are conveyed to the salivary
glands by the VII and IX cranial nerves.

Objectives   8; I; 4; (4.2–4.4).   8; IV; 3; (3.7), (3.9). 0.53   76   D
             8; VIII; 1; (1.4).   0.42   76

289   Lesions of the oculomotor nerve result in a divergent squint

BECAUSE

the *only* extrinsic ocular muscle not paralysed on the affected side is the lateral rectus.

Objectives   8; II; 1; (1.13), (1.14).   8; VIII; 1; (1.6). −   −   C

Type K items (Hubbard and Clemans, 1961) (Questions 290–354)

These items are in the form of incomplete statements and call for recognition of one or more correct responses which are grouped in various ways. The instruction for these items is as follows:

For each of the incomplete statements below, one or more of the completions given is correct. Responses should be chosen as follows:
(A) if only completions (i), (ii) and (iii) are correct,
(B) if only completions (i) and (iii) are correct,
(C) if only completions (ii) and (iv) are correct,
(D) if only completion (iv) is correct,
(E) if all completions are correct.

K(i)   Cranio-cerebral topography, eye, ear and nose (Questions 290–308)

290   Features of the sphenoid bone include

(i)   anterior clinoid processes.
(ii)   lateral pterygoid plates.
(iii) medial pterygoid plates.
(iv)   optic canals.

Objectives   8; I; 1; (1.1), (1.2), (1.5).   *0.54*   78   E

291   The hypophysial fossa

(i)   forms part of the middle cranial fossa.
(ii)   lies above the body of the sphenoid bone.
(iii) is readily visible in lateral radiographs of the skull.
(iv)   is closely related to the sphenoidal sinus.

Objectives   8; I; (1.1), (1.2), (1.7).   8; III; 1; (1.4).   *0.55*   86   E

292 Air cells (sinuses) are usually found in the

    (i) frontal bone.
    (ii) temporal bone.
    (iii) maxilla.
    (iv) parietal bone.

Objectives   8; III; 1; (1.4).   8; VII; 1; (1.2).   *0.32*   42   A
                                                        0.40   58
                                                        0.29   52

293 The first dentition

    (i) usually does not start to erupt before the age of
        one year.
    (ii) has a total of twenty teeth.
    (iii) contains two premolar teeth in each quadrant.
    (iv) contains two molar teeth in each quadrant.

Objectives   8; IV; 3; (3.12).   8; IX; 1; (1.1).   *0.23*   53   C
                                                         0.39   66

294 The posterior cranial fossa

    (i) has a floor in a plane below that of the middle
        cranial fossa.
    (ii) contains part of the abducent nerves.
    (iii) contains the basilar artery.
    (iv) contains part of the internal carotid arteries.

Objectives   8; I; 1; (1.2), (1.4).   8; I; 3; (3.3).   –   –   A

295 The jugular foramen transmits the

    (i) internal carotid artery.
    (ii) vagus nerve.
    (iii) hypoglossal nerve.
    (iv) internal jugular vein.

Objectives   8; I; 1; (1.4–1.6).
                 8; I; 3; (3.1), (3.3), (3.10).   *0.53*   90   C

296   The frontal bone

(i)   forms part of the floor of the middle cranial fossa.
(ii) forms part of the roof of the orbit.
(iii) forms the superior boundary of the superior
      orbital fissure.
(iv) contains two of the paranasal air sinuses.

Objectives   8; I; 1; (1.1), (1.2), (1.5).
             8; II. 1. (1.11).   8; III; 1; (1.4).   –   –   C

297   The zygomatic bone extends between the

(i)   maxilla and temporal bones.
(ii) frontal and temporal bones.
(iii) maxilla and sphenoid bones.
(iv) maxilla and frontal bones.

Objectives   8; I; 1; (1.1), (1.8).   –   –   E

298   The superior orbital fissure transmits

(i)   branches of the oculomotor nerve.
(ii) the abducent nerve.
(iii) branches of the ophthalmic nerve.
(iv) the ophthalmic artery.

Objectives   8; I; 1; (1.5), (1.6).   8; I; 3; (3.5).
             8; VIII; 1; (1.1), (1.5).   –   –   A

299   The ethmoidal cells

(i)   are situated medial to the lacrimal sac.
(ii) drain into both superior and middle nasal
      meatuses.
(iii) are situated between the lateral wall of the
      nose and medial wall of the orbit.
(iv) cannot be visualized in radiographs.

Objectives   8; I; 1; (1.1), (1.7).
             8; II; 1; (1.7), (1.11).
             8; III; 1; (1.1–1.5).   –   –   A

300   The bones forming the walls of the orbit include the

(i)   maxilla.
(ii)   sphenoid.
(iii) frontal.
(iv)   zygomatic.

Objectives   8; I; 1; (1.1), (1.8).   8; II; 1; (1.11).   —   —   E

301   The sphenoidal air sinus is situated

(i)   anterior to the pons.
(ii)   inferior to the optic chiasma.
(iii) inferior to the hypophysis cerebri.
(iv)   medial to the internal carotid artery.

Objectives   8; I; 1; (1.1–1.3).   8; I; 3; (3.1), (3.5).
8; III; 1; (1.4).   0.60   71   E

302   The maxillary sinus

(i)   has a lining membrane continuous with that of
the nose.
(ii)   drains into the middle meatus.
(iii) drains by an opening situated above the level
of its floor.
(iv)   occupies the body of the maxilla.

Objectives   8; I; 1; (1.8).   8; III; 1; (1.2), (1.4).   0.38   91   E

303   In the nasal cavity the

(i)   vestibule is below the atrium.
(ii)   olfactory mucosa is limited to the septum.
(iii) spheno-ethmoidal recess is postero-superior
to the superior meatus.
(iv)   cribiform plate is in the posterior part of the
roof.

Objectives   8; III; 1; (1.1), (1.2), (1.4), (1.6).   —   —   B

304   The ethmoidal sinuses open into the

(i)   superior meatus.
(ii) spheno-ethmoidal recess.
(iii) middle meatus.
(iv) inferior meatus.

Objectives   8; III; 1; (1.2–1.4).   –   –   B

305   The boundaries of the piriform aperture of the nasal
      cavity include

(i)   zygomatic bones.
(ii) maxillary bones.
(iii) frontal bones.
(iv) nasal bones.

Objectives   8; I; 1; (1.1), (1.8).
             8; III; 1; (1.1), (1.2).   –   –   C

306   Opening into hiatus semilunaris are the

(i)   posterior ethmoidal sinus.
(ii) maxillary sinus.
(iii) sphenoidal sinus.
(iv) frontal sinus.

Objectives   8; III; 1; (1.2–1.4).   –   –   C

307   The lacrimal gland

(i)   is situated antero-laterally near the roof of the
      orbit.
(ii) projects into the upper eyelid.
(iii) receives both sympathetic and parasympathetic
      innervation.
(iv) receives secreto-motor fibres from cranial nerve
      VII.

Objectives   8; I; 4; (4.2), (4.4).
             8; II; 1; (1.7), (1.15).   *0.25*   56   E

308 The semicircular canals

    (i)   are concerned with equilibration.
    (ii) are concerned with hearing.
    (iii) contain both endolymph and perilymph.
    (iv) can be examined with an auriscope.

Objectives   8; VII; 1; (1.4), (1.5), (1.7).   *0.21*   87   B
                                                   0.31   93
                                                 0.51   95

K(ii) Musculo-skeletal structures in the neck (Questions 309–311)

309 The atlas

    (i)   forms an ellipsoidal articulation with the skull.
    (ii) has transverse processes palpable between the
          mastoid process and the angle of the mandible.
    (iii) is the widest cervical vertebra.
    (iv) is involved in both nodding and rotational
          movements.

Objectives   8; I; 2; (2.1), (2.7).   *0.35*   56   E

310 The sternocleidomastoid muscle

    (i)   is supplied by the accessory nerve and the 2nd
          and 3rd cervical nerves.
    (ii) flexes the neck and rotates the face to the
          opposite side.
    (iii) has two heads.
    (iv) lies medial to the phrenic nerve.

Objectives   8; I; 2; (2.2), (2.3), (2.5), (2.7).   –   –   B

311 The sternocleidomastoid muscle

    (i)   rotates the head to the opposite side.
    (ii) tilts the head to the same side.
    (iii) may assist respiration.
    (iv) is supplied by the 1st and 2nd cervical nerves.

Objectives   8; I; 2.   *0.37*   58   A

K(iii) Blood vessels and lymph nodes (Questions 312–321)

312   The internal carotid artery

   (i)   enters the skull through the foramen lacerum.
   (ii) divides into the anterior and middle cerebral
         arteries.
   (iii) usually begins about the level of the 6th cervical
         vertebra.
   (iv) gives off the ophthalmic artery.

   Objectives   8; I; 3; (3.1), (3.3), (3.5).   *0.42*   89   C

313   The cervical portion of the internal carotid artery

   (i)   is comparatively superficial at its commencement.
   (ii) is palpable in the neck.
   (iii) begins at the level of the upper border of the
         thyroid cartilage.
   (iv) ascends in front of the upper 3 cervical vertebrae.

   Objectives   8; I; 3; (3.1), (3.2), (3.8).   8; VI; 1; (1.1). –   –   E

314   The internal carotid artery

   (i)   begins at the level of the upper margin of the
         thyroid cartilage.
   (ii) lies medial to the external carotid artery
         throughout its course.
   (iii) gives no branches in the neck.
   (iv) terminates in the anterior cranial fossa by
         dividing into the anterior and middle cerebral
         arteries.

   Objectives   8; I; 3; (3.1), (3.3), (3.5), (3.6).
                 8; VI; 1; (1.1).   *0.36*   44   B

315   Branches of the internal carotid artery include the

   (i)   middle meningeal artery.
   (ii) basilar artery.

(iii) infraorbital.artery.
(iv) ophthalmic artery.

Objectives   8; I; 3.   *0.45*   57   **D**

316 The external carotid artery supplies direct branches
     to the

(i)   tongue.
(ii) face.
(iii) thyroid gland.
(iv) brain.

Objectives   8; I; 3; (3.1), (3.2), (3.6).   *0.45*   87   **A**
                                                    0.67   93

317 The carotid sheath encloses the

(i)   internal jugular vein.
(ii) vagus nerve.
(iii) common carotid artery.
(iv) cervical sympathetic trunk.

Objectives   8; I; 3; (3.1), (3.3), (3.10).
              8; I; 4; (4.3).   8; VIII; 1; (1.1). *0.23*   83   **A**

318 The vertebral artery

(i)   arises from the subclavian artery.
(ii) enters a foramen in the transverse process of the
      6th cervical vertebra.
(iii) turns medially above the atlas.
(iv) terminates at the level of the foramen magnum
      by joining the artery of the opposite side.

Objectives   8; I; 1; (1.5).   8; I; 2; (2.1).   *0.29*   46   **A**
              8; I; 3; (3.1), (3.3).   *0.41*   63
                                                    0.34   39
                                                    0.53   74

319   The left subclavian artery

(i)   arises from the brachiocephalic trunk.
(ii) gives off the left vertebral artery.
(iii) becomes the axillary artery at the outer border
      of scalenus anterior.
(iv) can be compressed against the first rib.

Objectives   8; I; 3; (3.1).   *0.46*   75   C

320   The internal jugular veins drain blood from the

(i)   brain.
(ii) face.
(iii) tongue.
(iv) thyroid gland.

Objectives   8; I; 3; (3.10), (3.11).   —   —   E

321   The deep cervical group of lymph nodes is situated

(i)   along the course of the external jugular vein.
(ii) behind the carotid sheath.
(iii) superficial to the investing layer of the deep
      cervical fascia.
(iv) along the course of the internal jugular vein.

Objectives   8; I; 3; (3.1), (3.10).   *0.26*   50   D
                                                0.52   49

K(iv) Masticatory process, swallowing and phonation (Questions 322—334)

322   The function of the medial pterygoid muscle is to
      assist other muscles in

(i)   elevating the jaw.
(ii) retracting the jaw.
(iii) protruding the jaw.
(iv) moving the jaw laterally to the ipsilateral side.

Objectives   8; IV; 2; (2.1), (2.4), (2.5).   —   —   B

323 The medial pterygoid muscle

(i)   is innervated by the mandibular nerve.
(ii) is easily palpable.
(iii) is active during lateral excursion of the mandible.
(iv) has an attachment to the temporomandibular
      articular disc.

Objectives   8; IV; 1; (1.2).   8; IV; 2; (2.1–2.5).   *0.56*   87   B
                                                          0.71   93
                                                          0.43   96

324 Muscles that close the jaw include the

(i)   temporalis.
(ii) buccinator.
(iii) medial pterygoid.
(iv) lateral pterygoid.

Objectives   8; I; 1; (1.10).
             8; IV; 2; (2.1), (2.4), (2.5)   –   –   B

325 The neck of the mandible

(i)   is covered by the masseter muscle.
(ii) gives attachment to the lateral ligament of the
      temporomandibular joint.
(iii) gives attachment to the sphenomandibular
      ligament.
(iv) is covered by the parotid gland.

Objectives   8; I; 1; (1.9).   8; IV; 1; (1.1), (1.2).
             8; IV; 2; (2.1).   8; IV; 3; (3.7), (3.8).   *0.50*   53   C

326 The dorsum of the tongue posterior to the sulcus
     terminalis carries

(i)   fungiform papillae.
(ii) vallate papillae.
(iii) filiform papillae.
(iv) lymphoid nodules.

Objectives   8; IV; 3; (3.1).   *0.34*   44   D

327   The submandibular salivary gland

(i)   lies mainly superficial to the mylohyoid muscle.
(ii) lies deep to the hyoglossus muscle.
(iii) lies partly under cover of the mandible.
(iv) receives secretomotor (parasympathetic) supply
      from the glossopharyngeal nerve.

Objectives   8; IV; 2; (2.1).   8; IV; 3; (3.7), (3.9).
              8; VIII; 1; (1.4).   *0.54*   85   B

328   The lateral wall of the nasopharynx exhibits the

(i)   pharyngeal (nasopharyngeal) tonsil.
(ii) opening of the auditory tube.
(iii) palatopharyngeal fold.
(iv) pharyngeal recess.

Objectives   8; V; 1; (1.1), (1.6).   8; VII; 1; (1.2).   *0.21*   50   C
                                                             0.23   63
                                                             0.65   82

329   Boundaries of the oropharyngeal isthmus include the

(i)   soft palate.
(ii) tongue.
(iii) palatoglossal folds.
(iv) palatopharyngeal folds.

Objectives   8; IV; 3; (3.1), (3.3).   8; V; 1; (1.1).   –   –   A

330   The laryngeal inlet

(i)   is bounded by the vocal and vestibular folds.
(ii) is closed by adduction of the vocal folds.
(iii) faces mainly anteriorly.
(iv) has a piriform recess of the laryngopharynx on
      each side.

Objectives   8; V; 1; (1.1), (1.5).
              8; VI; 1; (1.1), (1.3), (1.4).   *0.35*   78   D

331 The vocal ligament

    (i)   is attached to the thyroid cartilage.
    (ii) is attached to the cricoid cartilage.
    (iii) is attached to the arytenoid cartilage.
    (iv) lies within the free edge of the vestibular fold.

Objectives 8; VI; 1; (1.1–1.4).   *0.44* 77   B

332 The thyroid cartilage

    (i)   has two laminae joined anteriorly at an angle.
    (ii) is attached to the hyoid bone by a membrane.
    (iii) forms part of the lateral boundary of the piriform
         fossa.
    (iv) does not articulate with the cricoid cartilage.

Objectives 8; V; 1; (1.5).   8; VI; 1; (1.1–1.3).   *0.44* 85   A

333 Normal voice and speech mechanisms

    (i)   involve the recurrent laryngeal nerves.
    (ii) require intrinsic laryngeal muscle activity.
    (iii) require intact hypoglossal nerves.
    (iv) involve the facial nerves.

Objectives 8; VI; 1; (1.5), (1.6).   *0.28* 77   E
               8; VIII; 1; (1.5), (1.6).   *0.56* 86
                                           0.50 87

334 The parotid gland contains

    (i)   parotid lymph nodes.
    (ii) the facial nerve.
    (iii) the external carotid artery.
    (iv) the mandibular nerve.

Objectives 8; IV; 3; (3.8).   –   –   A

K(v) Cranial nerves (Questions 335—354)

335   The oculomotor nerve

   (i)   emerges from the mid-brain.
   (ii) may be damaged resulting in ptosis.
   (iii) contains pre-ganglionic parasympathetic fibres.
   (iv) contains motor fibres to all the extrinsic eye
         muscles except the superior oblique.

   Objectives   8; I; 1; (1.3), (1.4).   8; I; 4; (4.2—4.4).
                8; II; 1; (1.13), (1.14).
                8; VIII; 1; (1.4—1.6).   *0.45*   80   A

336   The oculomotor nerve supplies the

   (i)   ciliary muscle.
   (ii) superior oblique muscle.
   (iii) inferior oblique muscle.
   (iv) conjunctiva.

   Objectives   8; II; 1; (1.1), (1.13).   —   —   B

337   The trigeminal nerve

   (i)   has a small motor and large sensory root.
   (ii) includes the nasal cavity in its sensory
         distribution.
   (iii) emerges from the lateral part of the pons.
   (iv) has a ganglion in the posterior cranial fossa.

   Objectives   8; I; 1; (1.2—1.4), (1.12).   8; III; 1.   —   —   A

338   The trigeminal nerve has branches traversing the

   (i)   foramen rotundum.
   (ii) superior orbital fissure.
   (iii) foramen ovale.
   (iv) foramen lacerum.

   Objectives   8; I; 1; (1.4), (1.5).   —   —   A

339 The fifth cranial nerve

    (i)   arises from the pons.
    (ii) is the motor nerve to the muscles of mastication.
    (iii) is sensory to the face.
    (iv) provides a sensory and a motor root to the
         semilunar ganglion.

Objectives   8; I; 1; (1.3), (1.4), (1.12).
                 8; IV; 2; (2.2).   –   –   A

340 Division of the mandibular nerve at its exit from the
     skull would result in

    (i)   sensory impairment over the anterior two-thirds
         of the tongue.
    (ii) paralysis of the pterygoid muscles.
    (iii) paralysis of the temporalis muscle.
    (iv) impairment of palatal function.

Objectives   8; I; 1; (1.4), (1.5).   8; IV; 2; (2.1), (2.2).
                 8; IV; 3; (3.4), (3.5).   *0.34*   35   E

341 The maxillary nerve

    (i)   leaves the skull through the foramen spinosum.
    (ii) enters the orbit through the superior orbital
         fissure.
    (iii) supplies the skin of the lower lip.
    (iv) supplies the mucous membrane of the nasal cavity.

Objectives   8; I; 1; (1.4–1.6), (1.12).   8; III; 1.   –   –   D

342 The roots of the fifth cranial nerve contain

    (i)   sensory fibres supplying the skin of the scalp.
    (ii) secretory fibres to the lacrimal gland.
    (iii) motor fibres to the masticatory muscles.
    (iv) motor fibres to the pharyngeal muscles.

Objectives   8; I; 1; (1.4), (1.12).   8; I; 4; (4.4).
                 8; II; 1; (1.8).   8; IV; 2; (2.1–2.3).
                 8; IV; 2; (2.1–2.3).   8; V; 1; (1.3).   *0.42*   87   B

343 If the maxillary nerve is severed at the foramen
     rotundum, the results include

   (i)   paralysis of the muscles of mastication.
   (ii) an area of loss of sensation over the forehead.
   (iii) loss of the corneal reflex.
   (iv) an area of loss of sensation over the temporal
         region.

Objectives   8; I; 1; (1.4–1.6), (1.12).
             8; II; 1; (1.8).   8; IV; 2; (2.1), (2.2).   *0.41*   98   **D**

344 The facial nerve

   (i)   emerges from the skull through the stylomastoid
         foramen.
   (ii) divides into upper and lower branches within the
         substance of the parotid gland.
   (iii) emerges from the parotid gland in 5 major
         divisions.
   (iv) contains cutaneous sensory fibres to the skin of
         the cheek.

Objectives   8; I; 1; (1.5), (1.6).   8; IV; 3; (3.7), (3.8).   –   –   **A**

345 The facial nerve contains

   (i)   motor fibres to skeletal muscle.
   (ii) special visceral afferent fibres.
   (iii) parasympathetic fibres.
   (iv) preganglionic sympathetic fibres.

Objectives   8; I; 4; (4.1–4.5).   8; VIII; 1; (1.2), (1.4).   *0.66*   76   **A**

346 Damage to the facial nerve may result in

   (i)   drooping of the mouth.
   (ii) inability to close the eyelids strongly.
   (iii) complete loss of the corneal reflex.
   (iv) complete loss of taste.

Objectives   8; I; 1; (1.10–1.14).   8; II; 1; (1.8).
             8; IV; 3; (3.4), (3.20).   8; VIII; 1; (1.6).   –   –   **A**

347 The glossopharyngeal nerve

   (i)   contains fibres associated with taste from the
         posterior third of the tongue.
   (ii) leaves the skull through the jugular foramen.
   (iii) has two sensory ganglia.
   (iv) is entirely sensory.

   Objectives   8; I; 1; (1.4–1.6).   8; IV; 3; (3.4).
                8; VIII; 1; (1.2).   –   –   A

348 The left recurrent laryngeal nerve

   (i)   supplies the mucous membrane of the larynx
         below the vocal folds.
   (ii) runs between the trachea and oesophagus in
         the neck.
   (iii) supplies part of the mucosa of the laryngeal
         pharynx.
   (iv) innervates the cricothyroid muscle.

   Objectives   8; V; 1.
                8; VI; 1.   0.32   42   A

349 In lesions of the external branch of the accessory
    nerve there is impaired ability to

   (i)   clench the teeth.
   (ii) rotate the face to the opposite side.
   (iii) depress the palate.
   (iv) elevate the scapula.

   Objectives   8; I; 2; (2.2–2.6).   8; IV; 2; (2.1), (2.2). 0.54   88   C
                8; IV; 3; (3.5).   8; VIII; 1; (1.6).   0.44   92

350 The hypoglossal nerve

   (i)   emerges from the medulla medial to the olive.
   (ii) emerges from the hindbrain lateral to the pyramid.
   (iii) leaves the skull through a canal (foramen) in the
         occipital bone.

(iv) supplies intrinsic but not extrinsic muscles of
the tongue.

Objectives   8; I; 1; (1.3–1.6).   —   —   A
8; IV; 3; (3.2), (3.5), (3.6).
8; VIII; 1; (1.5).

351   The hypoglossal nerve passes

(i) superficial to the posterior belly of the digastric
muscle.
(ii) superficial to the internal and external carotid
arteries.
(iii) superficial to the internal carotid artery but deep
to the external carotid artery.
(iv) between the internal jugular vein and the internal
carotid artery.

Objectives   8; I; 2; (2.2).   8; I; 3; (3.1), (3.3), (3.10).   —   —   C

352   The distribution of the phrenic nerve includes

(i) diaphragmatic parietal pleura.
(ii) diaphragmatic parietal peritoneum.
(iii) diaphragm.
(iv) lower intercostal muscles.

Objectives   8; I; 2.   —   —   A

353.   Preganglionic parasympathetic fibres

(i) are found in the oculomotor nerve.
(ii) synapse in the middle cervical ganglion.
(iii) release acetylcholine.
(iv) innervate the pupillary constrictor muscle.

Objectives   8; I; 4; (4.2–4.5).   8; II; 1; (1.5).   0.55   45   B
0.60   63
0.58   71

354 The cervical sympathetic trunk

    (i)   has no preganglionic fibres originating in the neck.
    (ii) is connected to cervical nerves by white rami.
    (iii) is connected to cervical nerves by grey rami.
    (iv) innervates the sphincter pupillae.

Objectives   8; I; 4; (4.2), (4.3).   8; II; 1; (1.5).   −   −   B

Chapter 4

Musculo-skeletal Trunk

Type A items (Hubbard and Clemans, 1961) (Questions 355–387)

These involve choosing one correct answer from five available choices. The instruction for these items is as follows:
Each of the incomplete statements below is followed by five suggested answers or completions. Select the one which is best in each case.

A(i) Vertebral column (Questions 355–360)

355 A typical thoracic vertebra

(A) has well marked foramina transversaria traversing its pedicles.
(B) articulates with ribs of the same number and the ribs above.
(C) has joints which contribute to rotation of the vertebral column.
(D) articulates at plane synovial and primary (hyaline) cartilaginous joints.
(E) has none of the above properties.

Objectives   9; I; 1; (1.1).   9; I; 2; (2.2), (2.3), (2.5).
              9; II; 1; (1.3).   *0.33*   37   C

356   The thoracic vertebral column

(A) consists of ten vertebrae.
(B) has a primary curvature.
(C) permits no rotation.
(D) has foramina transversaria.
(E) surrounds the cauda equina in the child.

Objectives   9; I; 1; (1.2).   9; I; 2; (2.4–2.6).   *0.29*   93   **B**

357   The vertebral column

(A) has primary curvatures in the cervical and lumbar
    regions.
(B) has eight cervical vertebrae.
(C) has four lumbar vertebrae.
(D) is most mobile in the thoracic region.
(E) does not depend upon the erector spinae group
    of muscles for support in the resting erect
    posture.

Objectives   9; I; 1.   9; I; 2; (2.4), (2.5).
             9; IV; 1; (1.5).   *0.23*   49   **E**

358   The spinal cord

(A) gives direct attachment to mixed spinal nerves.
(B) may be injured by lumbar punctures between
    LV1 and LV2.
(C) is surrounded by dura mater which terminates
    at the level of LV2.
(D) gives rise to the roots of a total of seven pairs of
    cervical spinal nerves.
(E) is immediately surrounded by cerebrospinal fluid.

Objectives   9; I; 2; (2.6).   *0.59*   85   **B**

359   The spinal cord

(A) occupies the length of the vertebral canal.
(B) is immediately surrounded by arachnoid mater.

     (C) is immediately surrounded by pia mater.
     (D) is immediately surrounded by cerebrospinal fluid.
     (E) gives direct attachment to the mixed spinal nerves
        constituting the cauda equina.

Objectives   9; I; 2; (2.6).   *0.23*   90   **C**
                                            0.54   87

360   The intervertebral discs

     (A) form part of the anterior boundaries of
        intervertebral foramina.
     (B) account for approximately half the length of
        the newborn vertebral column.
     (C) constitute primary cartilaginous joints.
     (D) are more likely to rupture in old age because of
        water uptake.
     (E) produce the primary and secondary curvatures
        of the vertebral column.

Objectives   9; I; 2.   *0.46*   67   **A**

A(ii) Framework of thorax (Questions 361–364)

361   Ventral roots of the thoracic spinal nerves contain

     (A) somatic motor fibres.
     (B) somatic sensory fibres.
     (C) postganglionic visceral efferent fibres.
     (D) unmyelinated preganglionic visceral efferent
        fibres.
     (E) visceral afferent fibres.

Objectives   9; II; 3; (3.8).   *0.45*   84   **A**
                                              0.34   88

362   The sternal angle

     (A) marks the xiphisternal joint.
     (B) is usually at the level of the 3rd costal cartilage.
     (C) is situated above the jugular notch.

(D) forms part of the socket for the sternoclavicular
joint.
(E) varies during breathing.

Objectives   9; II; 1; (1.2), (1.4), (1.7).   0.43   77   E
                                                            0.70   91

363   In relation to the diaphragm the

(A) inferior vena cava passes through its muscular part.
(B) aorta passes behind it in the midline at the level
of the 10th thoracic vertebra.
(C) left crus has more extensive attachments than the
right.
(D) oesophagus passes through it at the level of the
8th thoracic vertebra.
(E) anterior and posterior vagal trunks pass through
the oesophageal opening.

Objectives   9; II; 2; (2.1−2.3).   0.41   83   E
                                                            0.26   72

364   The middle mediastinum contains the

(A) vagus nerves.
(B) arch of the aorta.
(C) descending aorta.
(D) heart.
(E) azygos vein.

Objectives   9; III; 3; (3.1), (3.2), (3.5).   0.58   80   D
                                                            0.56   93

A(iii) Framework of abdomen (Questions 365−376)

365   The external oblique muscle

(A) becomes aponeurotic at the arcuate line.
(B) has its fibres at right angles to those of the
opposite internal oblique.
(C) has no action on the vertebral column.

(D) has no action on the ribs.
(E) has none of the above properties.

Objectives   9; II; 1; (1.7), (1.14).   *0.52*   74   E

366   The external oblique muscle

(A) has no aponeurosis.
(B) forms the inguinal ligament.
(C) forms part of the posterior wall of the rectus
    sheath.
(D) has no attachment to the transversus abdominis
    muscle.
(E) has none of the above properties.

Objectives   9; III; 1; (1.7), (1.8).   *0.29*   72   B
                                                          0.29   75

367   The arcuate line of the rectus sheath is

(A) anterior to the rectus abdominis muscle.
(B) at the level of the umbilicus.
(C) lateral to the rectus abdominis muscle.
(D) the line below which the transversus aponeurosis
    passes anterior to the rectus abdominis muscle.
(E) none of the above.

Objectives   9; III; 1; (1.7).   *0.28*   63   D
                                                          0.31   63
                                                          0.36   81
                                                          0.47   71

368   The rectus abdominis muscle

(A) is supplied by lumbar nerves.
(B) is active in quiet expiration.
(C) forms part of the anterior wall of the inguinal
    canal.
(D) is divided by tendinous intersections mainly
    above the umbilicus.
(E) is attached to the linea alba.

Objectives   9; III; 1; (1.7), (1.9), (1.14).
              9; III; 3; (3.6), (3.8).   —   —   D

369 The rectus abdominis muscle

(A) contracts only during resisted flexion of the
    trunk.
(B) is supplied by T7—12 ventral rami.
(C) is supplied wholly by the inferior epigastric
    artery.
(D) is completely enclosed within the internal
    oblique aponeurosis.
(E) has none of the above properties.

Objectives   9; III; 1; (1.7).   9; III; 3; (3.2), (3.8).   –   –   B

370 The rectus abdominis muscle

(A) usually has 3 tendinous intersections attached to
    the anterior and posterior walls of the rectus
    sheath.
(B) is supplied by T7—12 ventral rami.
(C) is completely enclosed within the internal oblique
    aponeurosis.
(D) does not extend above the costal margin.
(E) should be retracted medially in an abdominal
    incision to avoid paralysis.

Objectives   9; III; 1.   0.30   68   B

371 The deep inguinal ring

(A) is medial to the conjoint tendon.
(B) transmits the ductus deferens.
(C) is an opening in the transversus abdominis muscle.
(D) is situated about 1 cm medial to the anterior
    superior iliac spine.
(E) forms the neck of a direct inguinal hernia.

Objectives   9; III; 1; (1.9), (1.10).   0.41   80   B
                                                 0.40   47

372   The conjoint tendon (falx inguinalis) is formed by
      contributions from the

      (A) internal oblique and external oblique muscles.
      (B) transversus abdominis muscle only.
      (C) cremaster muscle and transversus abdominis
          muscle.
      (D) transversus abdominis and internal oblique
          muscles.
      (E) external spermatic fascia and internal oblique
          muscle.

      Objectives   9; III; 1.   *0.37* 81   D
                                                 0.31 88

373   The inguinal ligament

      (A) is the lower border of the internal oblique
          aponeurosis.
      (B) is attached to the pubic tubercle.
      (C) forms the roof of the inguinal canal.
      (D) is above the superficial inguinal ring.
      (E) has none of the above properties.

      Objectives   9; III; 1; (1.8), (1.9).   *0.54* 96   B

374   The inguinal canal

      (A) has the internal oblique muscle in the lateral
          part of its posterior wall.
      (B) is situated below the inguinal ligament.
      (C) has the transversus abdominis in the medial part
          of its anterior wall.
      (D) has the internal oblique muscle in the lateral
          part of its anterior wall.
      (E) has an external ring directly lateral to the pubic
          tubercle.

      Objectives   9; III; 1; (1.9).   *0.56* 64   D
                                                 0.47 86

375   The superficial inguinal ring

(A) is an opening in the transversalis fascia.
(B) lies directly lateral to the pubic tubercle.
(C) is strengthened in front by the conjoint tendon.
(D) has the pubic crest at its base.
(E) has none of the above properties.

Objectives   9; III; 1; (1.9).   —   —   D

376   The psoas major and quadratus lumborum muscles
      both

(A) flex the trunk at the hip.
(B) flex the lumbar vertebrae laterally.
(C) are important in maintaining the erect posture.
(D) flex the lumbar vertebrae forwards.
(E) stabilize the ribs in respiration.

Objectives   9; III; 1; (1.11), (1.17).   0.30   31   B
                                                 0.34   49

A(iv) Framework of pelvis (Questions 377–382)

377   The sacrum

(A) articulates with the fourth lumbar vertebra.
(B) is attached by ligaments to both the ilium and
    ischium.
(C) is broader in the male than in the female.
(D) lies partly above the inlet of the true pelvis.
(E) has three segments.

Objectives   9; I; 1; (1.2), (1.3).
             9; III; 2; (2.1), (2.2), (2.4).   —   —   B

378   The sacrum

(A) usually has six segments.
(B) articulates with the fourth lumbar vertebra.
(C) is crossed by the obturator nerves.

(D) is broader behind than in front.
(E) has none of the above properties.

Objectives   9; I; 1; (1.3).   9; III; 2; (2.1), (2.4).
          9; III; 3; (3.7).   –   –   C

379   The pubic bones

(A) form an angle that is less acute in the male than
    in the female.
(B) give attachment to inguinal and lacunar ligaments.
(C) have superior rami which meet at the pubic
    symphysis.
(D) unite at a primary cartilaginous joint.
(E) have none of the above properties.

Objectives   9; III; 1; (1.2), (1.5), (1.8).
          9; III; 2; (2.1), (2.2).   0.34   88   B

380   The levator ani muscle

(A) forms the urogenital diaphragm.
(B) lies posterior to the ischiorectal fossa.
(C) forms part of the pelvic diaphragm.
(D) lies superior to the pelvic diaphragm.
(E) forms the lateral wall of the ischiorectal fossa.

Objectives   9; III; 2; (2.7), (2.10), (2.12).   0.66   89   C
                                                        0.58   88

381   The levatores ani muscles

(A) form the lateral walls of the ischiorectal fossae.
(B) constitute the urogenital diaphragm.
(C) are important in continence of urine and faeces.
(D) are not attached to the central tendon of the
    perineum.
(E) have none of the above properties.

Objectives   9; III; 2; (2.7), (2.8), (2.10–2.12).   –   –   C

382   The urogenital diaphragm

(A) is innervated by cord segments S2, S3 and S4
      via the pudendal nerves.
(B) lies superior to the pelvic diaphragm.
(C) is in a horizontal plane in the erect subject.
(D) is limited posteriorly by a line just posterior to
      the ischial tuberosities.
(E) is limited laterally by the ischiorectal fossae.

Objectives   9; III; 2; (2.10), (2.12).   9; III; 3; (3.7).   *0.40*   68   A

A(v) Blood vessels and nerves (Questions 383–387

383   The inferior vena cava

(A) is formed by junction of the internal iliac veins.
(B) is formed at the level of the 4th lumbar vertebra.
(C) is formed at the level of the transtubercular plane.
(D) pierces the muscular part of the right cupola of
      the diaphragm.
(E) has none of the above properties.

Objectives   9; III; 3; (3.3), (3.4).   —   —   C

384   The external iliac arteries

(A) are larger than the internal iliac arteries.
(B) are medial to the external iliac veins.
(C) commence at the level of the supracristal plane.
(D) have their origins indicated on the surface by the
      lower point of trisection of a line between the
      anterior superior iliac spine and the pubic
      symphysis.
(E) have none of the above properties.

Objectives   9; III, 3; (3.1), (3.4).   *0.35*   61   A

385   The pudendal nerve

(A) contains postganglionic parasympathetic fibres.
(B) runs in the medial wall of the ischiorectal fossa.
(C) leaves the pelvis through the greater sciatic
      foramen.

(D) arises from S4 and S5.

(E) is purely motor to the perineum.

Objectives   9; III; 3; (3.6), (3.7).   –   –   C

386   The lumbosacral plexus

(A) is an autonomic plexus.

(B) is made up of the dorsal rami of L2 to S4.

(C) gives rise to the obturator nerve.

(D) incorporates a lumbosacral trunk usually
derived from L5 and S1.

(E) has none of the above properties.

Objectives   9; III; 3; (3.6).   0.34   58   C

387   The branch of the lumbosacral plexus which appears
at the medial border of the psoas major muscle is the

(A) iliohypogastric nerve.

(B) sciatic nerve.

(C) femoral nerve.

(D) obturator nerve.

(E) genitofemoral nerve.

Objectives   9; III; 3; (3.7).   0.32   56   D
                                                                 0.52   65
                                                                 0.56   93

Type E items (Hubbard and Clemans, 1961) (Questions 388–404)

These items are of the assertion—reason type and there are five possible
responses. The instruction for these items is as follows:

*Each question consists of an assertion and a reason. Responses should be
chosen as follows:*

*(A) if the assertion and reason are true statements and the reason is a
correct explanation of the assertion;*

*(B) if the assertion and reason are true statements but the reason is
NOT a correct explanation of the assertion;*

(C) if the assertion is true but the reason is a false statement;

(D) if the assertion is false but the reason is a true statement;

(E) if both assertion and reason are false statements.

E(i) Vertebral column (Questions 388–390)

388   Backward protrusion of a lumbar invertebral disc
tends to occur in the flexed position under load

BECAUSE

the erector spinae muscle is inactive throughout
flexion of the lumbar vertebral column.

Objectives   9; I; 2; (2.1), (2.5).   9; IV; 1; (1.5).   *0.43*   93   C

389   The erector spinae is contracting strongly just before
a weight is placed on the ground by a person standing
with knees extended

BECAUSE

the erector spinae controls spinal flexion throughout
its range.

Objectives   9; IV; 1; (1.5).   —   —   D

390   The joints between vertebral bodies are synovial joints

BECAUSE

the vertebral bodies are superiorly and inferiorly
covered by plates of hyaline cartilage.

Objectives   9; I; 2; (2.1).   *0.35*   64   D

E(ii) Framework of thorax (Questions 391–395)

391   The manubriosternal joint is immovable

BECAUSE

it contains a disc of fibrocartilage.

Objectives   9; II; 1; (1.4).   *0.47*   41   D
                                    0.31   75
                                    0.37   71

392 Expansion of the lower thorax in inspiration occurs
mainly in the antero-posterior diameter

BECAUSE

it depends chiefly on 'pump-handle' rotation of the
ribs.

Objectives   9; II; 1; (1.5).   *0.51*   39   E

393 Muscles of the antero-lateral abdominal wall are
active during forced but not quiet expiration

BECAUSE

there is no abdominal movement in quiet respiration.

Objectives   9; II; 1; (1.8).   9; III; 1; (1.14).   *0.23*   58   C
                                                      0.36   71

394 The sympathetic trunk has only grey rami
communicantes in the thorax

BECAUSE

the sympathetic outflow is restricted to the
thoracic region.

Objectives   9; II; 3; (3.8).   *0.47*   60   E
                                                 0.44   90

395 Removal of the lumbar part of both sympathetic
trunks would effectively provide sympathetic
denervation within the abdomen

BECAUSE

all the pre-ganglionic sympathetic nerve fibres to
abdominal viscera are carried in ventral rami above
the L2 level.

Objectives   9; II; 3; (3.8).   *0.37*   28   D

E(iii) Framework of abdomen (Questions 396–398)

396   The rectus abdominis muscle is made more powerful
       by the presence of transverse intersections attached
       to anterior and posterior walls of its sheath

       BECAUSE

       the power of a muscle is dependent on the number of
       muscle fibres it contains and not on their length.

       Objectives   9; III; 1.   –   –   D

397   Transverse incisions in the anterior abdominal wall
       usually heal with a fine scar

       BECAUSE

       they run parallel to the skin creases and Langer's
       lines of skin cleavage.

       Objectives   9; III; 1.   *0.28*   91   A

398   It is structurally advantageous that the conjoint
       tendon reinforces the inguinal canal in its medial part

       BECAUSE

       the anterior wall of the medial part of the canal is
       weakened at the site of the deep inguinal ring.

       Objectives   9; III; 1;   *0.49*   53   C
                                                            0.60   83
                                                            0.52   80

E(iv) Framework of pelvis (Questions 399–404)

399   The ischiorectal fossa extends anteriorly above the
       urogenital diaphragm

       BECAUSE

       it extends forwards below the levator ani which has
       a part attached anteriorly to the body of the pubic
       bone.

       Objectives   9; III; 2; (2.7), (2.10), (2.12).   –   –   A

400   The bony pelvis is immovable

BECAUSE

the sacro-iliac joint and the pubic symphysis are
strong fibrous joints.

Objectives   9; III; 2; (2.1).   *0.43*   19   E

401   Relaxation of the puborectalis is important in
defaecation

BECAUSE

this relaxation helps to straighten the angle of
the perineal flexure.

Objectives   9; III; 2; (2.8).   *0.25*   61   A

402   During parturition (birth) the fetal head occupies
the oblique diameter of the true pelvic cavity in
the gynaecoid type of pelvis

BECAUSE

the distance between the ischial tuberosities is
greater in the gynaecoid than in the android type
of pelvis.

Objectives   9; III; 2.   *0.49*   67   B

403   The fetal head engages in the antero-posterior
(conjugate) diameter of the pelvic brim

BECAUSE

the maximum diameter of the fetal head is
transverse.

Objectives   9; III; 2.   -   *0.38*   41   E
                                                        0.53   49

404   The sacrotuberous ligament runs between the sacrum
and the ilium

BECAUSE

its principal role is to resist rotation at the sacro-iliac
joint.

Objectives   9; III; 2; (2.2), (2.3).   *0.51*   51   D

Type K items (Hubbard and Clemans, 1961) (Questions 405–439)

These items are in the form of incomplete statements and call for recognition
of one or more correct responses which are grouped in various ways. The
instruction for these items is as follows:

For each of the incomplete statements below, one or more of the
completions given is correct. Responses should be chosen as follows:
(A) if only completions (i), (ii) and (iii) are correct,
(B) if only completions (i) and (iii) are correct,
(C) if only completions (ii) and (iv) are correct,
(D) if only completion (iv) is correct,
(E) if all completions are correct.

K(i) Vertebral column (Questions 405–410)

405   A 'typical' cervical vertebra has

(i)   a relatively small vertebral foramen.
(ii) a bifid spinous process.
(iii) a relatively large body.
(iv) foramina transversaria.

Objectives   9; I; 1; (1.2), (1.4).   *0.36*   76   C
                                                     0.58   85
                                                     0.52   94

406   A 'typical' cervical vertebra has

(i)   a bifid spinous process.
(ii) a relatively small body.
(iii) foramina transversaria.
(iv) superior articular (zygapophyseal) processes
with anteriorly directed articular surfaces.

Objectives   9; I; 1; (1.2), (1.4).   *0.28*   59   A

407   The vertebral canal

(i)   is continued into the sacrum.
(ii) contains the lumbar spinal ganglia.
(iii) contains a vertebral venous plexus.
(iv) does not extend below L1 in the adult.

Objectives   9; l; 1; (1.3).   9; l; 2; (2.6), (2.7).   *0.16*   67   B

408   The vertebral canal

(i)   is bounded antero-laterally by the vertebral
      laminae.
(ii) terminates at the level of the sacral promontory.
(iii) contains the anterior longitudinal ligament.
(iv) is narrowed in the thoracic region.

Objectives   9; l; 1; (1.1–1.3).   9; l; 2; (2.2).   *0.53*   81   D

409   The epidural space

(i)   contains cerebrospinal fluid.
(ii) contains the vertebral venous plexus.
(iii) contains the ligamenta denticulata.
(iv) continues into the sacral canal.

Objectives   9; l; 2; (2.6), (2.7).   *0.44*   76   C
                                                           0.49   70

410   The epidural space

(i)   is traversed during lumbar puncture.
(ii) contains the vertebral venous plexus.
(iii) contains a quantity of semi-fluid fat.
(iv) does not continue into the cranial cavity.

Objectives   9; l; 2.   *0.23*   25   E

K(ii) Framework of thorax (Questions 411–418)

411 Blood flow in the azygos vein

   (i)   drains into the superior vena cava.
   (ii) is connected with that in the vertebral venous
         plexus.
   (iii) drains the body wall.
   (iv) is phasically influenced by respiratory movement.

   Objectives   9; II; 3; (3.5).   *0.38*   66   E

412 The aortic opening in the diaphragm usually

   (i)   transmits the splanchnic nerves.
   (ii) lies between the crura.
   (iii) transmits the vagi.
   (iv) transmits the azygos vein.

   Objectives   9; II; 2; (2.3).   9; II; 3; (3.4), (3.7), (3.9).   –   –   C

413 The diaphragm

   (i)   has attachments to lumbar vertebral bodies by
         way of the crura.
   (ii) aids venous return.
   (iii) is innervated by the phrenic nerves.
   (iv) increases the transverse diameters of the lower
         thorax and upper abdomen by its contraction.

   Objectives   9; II; 2; (2.2), (2.4).   9; II; 3; (3.10).   *0.37*   80   E

414 The diaphragm

   (i)   is pierced through the central tendon by the
         oesophagus.
   (ii) is pierced through the central tendon by the
         inferior vena cava.
   (iii) is pierced through the central tendon by the aorta.
   (iv) receives motor supply solely from the phrenic
         nerves.

   Objectives   9; II; 2; (2.3).   9; II; 3; (3.10).   *0.50*   79   C

415   The costo-vertebral origins of the diaphragm are from
       the

    (i)   lower four ribs.
    (ii) lower six costal cartilages.
    (iii) 11th and 12th thoracic and 1st lumbar vertebrae.
    (iv) 1st three lumbar vertebrae.

Objectives   9; II; 2; (2.2).   –   –   C

416   The arch of the aorta

    (i)   lies behind the manubrium of the sternum.
    (ii) gives off the right subclavian artery.
    (iii) gives off the brachiocephalic trunk.
    (iv) gives off the right common carotid artery.

Objectives   9; II; 3; (3.2).   0.53   84   B

417   The approximate vertebral level of the

    (i)   jugular (sternal) notch is TV 3.
    (ii) sternal angle is TV 4.
    (iii) xiphisternal joint is TV 10.
    (iv) spine of the scapula is TV 3.

Objectives   9; II; 1; (1.6).   –   –   E

418   The second to sixth ribs each

    (i)   articulates with a costal cartilage.
    (ii) articulates with two vertebrae.
    (iii) has a costal groove on its internal surface.
    (iv) articulates with a transverse process.

Objectives   9; II; 1; (1.1).   0.36   93   E

K(iii) Framework of abdomen (Questions 419–426)

419   The muscles of the antero-lateral abdominal wall
      contract actively during

(i)   coughing.
(ii) quiet inspiration.
(iii) lateral flexion of the trunk.
(iv) forced inspiration.

Objectives   9; III; 1; (1.14).   0.39   78   B
                                                   0.40   77

420   The inguinal ligament

(i)   is attached to the anterior superior iliac spine.
(ii) is the edge of the aponeurosis of external oblique.
(iii) is attached to the pubic tubercle.
(iv) forms the floor of the inguinal canal.

Objectives   9; III; 1; (1.8), (1.9).   0.42   81   E

421   The trans-tubercular plane

(i)   is a horizontal plane through the highest points
      of the iliac crest.
(ii) is a horizontal plane through the pubic tubercles.
(iii) is at the level of LV4.
(iv) is at a distance of two vertebral bodies below
      the subcostal plane.

Objectives   9; III; 1; (1.13).   —   —   D

422   The transversus abdominis muscle forms

(i)   part of the posterior wall of the inguinal canal
      (in conjoint tendon).
(ii) part of the anterior rectus sheath.
(iii) part of the posterior rectus sheath.
(iv) the opening of the deep inguinal ring.

Objectives   9; III; 1; (1.7). (1.9).   0.52   73   A

423   The posterior abdominal wall contains

(i)   psoas major.
(ii) quadratus lumborum.
(iii) transversus abdominis.
(iv) diaphragm.

Objectives   9; III; 1; (1.11).   *0.55*   84   E

424   Indirect inguinal herniae

(i)   pass through the transversalis fascia.
(ii) frequently extend into the scrotum.
(iii) form a lump below and lateral to the pubic
      tubercle.
(iv) traverse the deep inguinal ring.

Objectives   9; III; 1; (1.10).   –   –   C

425   The rectus abdominis muscle

(i)   lies partly on transversalis fascia.
(ii) is surrounded throughout its extent by an
      aponeurotic sheath.
(iii) receives motor innervation from thoraco-
      abdominal nerves.
(iv) is normally active during standing.

Objectives   9; III; 1; (1.7), (1.14).   *0.37*   78   B
                                                         0.50   51

426   The psoas major muscle

(i)   lies anterior to the lumbar plexus.
(ii) flexes the trunk.
(iii) extends the thigh.
(iv) lies medial to the quadratus lumborum muscle.

Objectives   9; III; 1; (1.11).   9; III; 3; (3.6).   –   –   C

K(iv) Framework of pelvis (Questions 427—435)

427   The bony pelvis

    (i)   transmits the weight of the trunk to the lower
       limb.
    (ii) has comparatively wider diameters in the female
       than the male.
    (iii) affords protection to the pelvic viscera.
    (iv) consists entirely of the two hip bones.

Objectives   9; III; 2.   *0.28* 82   A
                                          0.34 95

428   The pelvic inlet or brim

    (i)   is on a horizontal plane.
    (ii) is proportionately larger in the female.
    (iii) is bounded laterally by the sacro-tuberous
       ligaments.
    (iv) is bounded posteriorly by the sacral promontory.

Objectives   9; III; 2; (2.2), (2.4).   *0.33* 74   C
                                          0.49 94

429   The tendinous centre of the perineum

    (i)   gives attachment to both muscle and fascia.
    (ii) lies directly anterior to the anal canal.
    (iii) is fused with the urogenital diaphragm.
    (iv) gives attachment to the levator prostatae.

Objectives   9; III; 2; (2.11).   *0.24* 48   E

430   The sacrospinous ligament

    (i)   forms a boundary of the lesser sciatic foramen.
    (ii) separates the greater and lesser sciatic foramina.
    (iii) has the coccygeus muscle on its pelvic aspect.
    (iv) contributes to the stability of the sacro-iliac joint.

Objectives   9; III; 2; (2.2), (2.3).   *0.41* 82   E

431   The pubococcygeus muscle is

    (i)   the most posterior part of the levator ani.
    (ii) inserted into the tendinous centre of the perineum.
    (iii) seldom damaged during parturition.
    (iv) the main part of the levator ani.

Objectives   9; III; 2; (2.7).   —   —   C

432   The piriformis muscle

    (i)   together with the levator ani, forms part of the
        pelvic floor.
    (ii) lies above the level of the ischial spine.
    (iii) has the sacral plexus superior to it.
    (iv) forms part of the posterior wall of the pelvis.

Objectives   9; III; 2; (2.9).   —   —   C

433   The obturator internus muscle forms

    (i)   part of the lateral pelvic wall.
    (ii) a boundary of the lesser sciatic foramen.
    (iii) a wall of the ischiorectal fossa.
    (iv) part of the pelvic floor.

Objectives   9; III; 2; (2.6).   —   —   B

434   The boundaries of the inferior aperture of the pelvis
     include the

    (i)   sacro-tuberous ligament.
    (ii) sacro-spinous ligament.
    (iii) inferior aspect of the pubic symphysis.
    (iv) posterior sacro-iliac ligament.

Objectives   9; III; 2.   *0.36*   85   B

435   The obturator internus muscle

    (i)   forms part of the lateral pelvic wall.
    (ii) passes through the lesser sciatic foramen.

(iii) forms part of the lateral wall of the ischiorectal
fossa.
(iv) is inferior to the levator ani.

Objectives   9; III; 2; (2.6).   —   —   A

K(v) Blood vessels and nerves (Questions 436–439)

436   The sympathetic trunk

(i)   is connected with sacral nerves via grey rami
communicantes.
(ii) is connected with thoraco-abdominal nerves via
grey and white rami communicantes.
(iii) lies medial to pelvic sacral foramina.
(iv) contains both preganglionic and postganglionic
fibres.

Objectives   9; III; 3; (3.9).   —   —   E

437   The lumbosacral plexus

(i)   gives direct branches to muscles of the posterior
abdominal wall.
(ii) includes lumbar roots from ventral rami of L2
to L5.
(iii) has a lumbosacral trunk consisting of the lower
part of the ventral ramus of the 4th lumbar
nerve and all of the ventral ramus of the 5th
lumbar nerve.
(iv) has its sacral part lying on the levator ani.

Objectives   9; III; 3; (3.6), (3.7).   *0.21*   22   A

438   The inferior vena cava

(i)   receives tributaries in the pelvis.
(ii) commences slightly below and to the right of
the termination of the right common iliac
artery.

(iii) commences at the level of the 1st sacral
vertebra, a little to the right of the median
plane.
(iv) is valveless.

Objectives   9; III; 3; (3.3).   —   —   C

439   The inferior vena cava

(i)   leaves the abdominal cavity at the level of the
8th thoracic vertebra.
(ii) is formed at the level of the 5th lumbar vertebra.
(iii) is posterior to the right common iliac artery.
(iv) pierces the central tendon of the diaphragm.

Objectives   9; III; 3; (3.3).   —   —   E

Chapter 5

The Lower Limb

Type A items (Hubbard and Clemans, 1961) (Questions 440–448)

These involve choosing one correct answer from five available choices. The instruction for these items is as follows:
Each of the incomplete statements below is followed by five suggested answers or completions. Select the one which is best in each case.

A(i) Hip region (Questions 440–448)

440 All the hamstring muscles

    (A) are medial rotators of the leg.
    (B) are extensors of both the thigh and the leg.
    (C) are innervated by the tibial component of the sciatic nerve.
    (D) are innervated by the peroneal component of the sciatic nerve.
    (E) actively contract in maintenance of the resting erect posture.

    Objectives   10; IV; 1.   –   –   C

441 The semimembranosus muscle

    (A) produces lateral rotation of the leg at the knee joint.

(B) helps to limit flexion at the hip when the knee is extended.
(C) has two heads of origin.
(D) is supplied by the common peroneal division of the sciatic nerve.
(E) crosses superficial to the sciatic nerve in the upper part of the thigh.

Objectives   10;I;2;(2.1);3;(3.6). 10;III;2;(2.5).   *0.51*   82   B

442   The opening in the adductor magnus muscle transmits the

(A) femoral vessels.
(B) femoral nerve.
(C) saphenous nerve.
(D) tibial nerve.
(E) sciatic nerve.

Objectives   10; I; 2.   *0.53*   81   A

443   The adductor muscles of the thigh are arranged antero-posteriorly in the order

(A) longus, magnus, brevis.
(B) brevis, longus, magnus.
(C) longus, brevis, magnus.
(D) brevis, magnus, longus.
(E) in none of the above listed orders.

Objectives   10; I; 2; (2.1).   *0.45*   90   C

444   The femoral triangle is bounded by

(A) the inguinal ligament, pectineus and sartorius muscles.
(B) the inguinal ligament, adductor longus and gracilis muscles.
(C) the inguinal ligament, rectus femoris and sartorius muscles.

(D) the inguinal ligament, adductor longus and
sartorius muscles.

(E) none of the above combinations of structure.

Objectives   10; 1; 2; (2.2).   *0.47*   96   D

445   Difficulty in abduction at the hip joint results from
paralysis of the

(A) gluteus maximus muscle.
(B) obturator externus muscle.
(C) vastus lateralis muscle.
(D) gluteus medius muscle.
(E) obturator internus muscle.

Objectives   10; 1; 2; (2.3).   *0.63*   94   D
                                                  0.63   98

446   Abduction at the hip joint

(A) depends upon the integrity of the obturator
nerve.
(B) may be produced by movements of the pelvis.
(C) results in apparent shortening of the ipsilateral
lower limb.
(D) depends upon the integrity of the inferior gluteal
nerve.
(E) is limited by the iliofemoral ligament.

Objectives   10; 1; 1.   *0.32*   46   B
                                                  0.28   38
                                                  0.46   77

447   Medial rotators of the thigh include the

(A) tensor fasciae latae muscle.
(B) obturator internus muscle.
(C) obturator externus muscle.
(D) piriformis muscle.
(E) sartorius muscle.

Objectives   10; 1; 2; (2.3).   *0.47*   68   A
                                                  0.38   77

448 Medial rotation of the thigh at the hip joint is
produced by

(A) the gluteus minimus muscle.
(B) the adductor longus muscle.
(C) the gracilis muscle.
(D) the pectineus muscle.
(E) none of the above muscles.

Objectives   10; I; 2; (2.3).   *0.60*   82   A

A(ii) Knee region (Questions 449–463)

449 The tibia

(A) articulates by its medial condyle with the fibula.
(B) has a medial malleolus projecting further distally
than the lateral malleolus of the fibula.
(C) is in direct contact with the common peroneal
nerve.
(D) articulates with the superior and medial surfaces
of the body of the talus.
(E) articulates with the articular facets of the patella.

Objectives   10; I; 3; (3.2).   10; II; 1; (1.1), (1.2), (1.5). *0.35*   76   D
                                                                 0.35   89
                                                                 0.38   87

450 The patella

(A) is in contact with femur and tibia.
(B) has an apex pointing superiorly.
(C) is enveloped on both surfaces by the tendon of
the quadriceps femoris muscle.
(D) has a layer of synovial membrane between it
and the femur.
(E) articulates with the femur but not the tibia.

Objectives   10; I; 3.   –   –   E

451   The suprapatellar bursa

(A) does not communicate with the knee joint.
(B) lies in front of the vastus intermedius muscle.
(C) extends 5 cm or more above the patella.
(D) may extend down into the infrapatellar fat pad.
(E) has none of the above properties.

Objectives   10; I; 3; (3.3).   *0.30*   66   C
                                                 0.46   81

452   The short head of the biceps femoris muscle

(A) acts during extension of the hip joint.
(B) is supplied by the tibial component of the
    sciatic nerve.
(C) is closely related to the vastus medialis.
(D) acts during flexion of the knee joint.
(E) has none of the above properties.

Objectives   10; I; 2; (2.1); 3; (3.6).
             10; III; 2; (2.5).   *0.36*   88   D

453   The semimembranosus muscle

(A) is an abductor at the hip joint.
(B) has two heads.
(C) is inserted into the fibula.
(D) is not used in normal walking along level ground.
(E) is a medial rotator of the tibia.

Objectives   10; I; 3.   *0.74*   89   E

454   The quadriceps femoris muscle

(A) aids in flexion at the hip joint.
(B) is supplied by the 5th lumbar segment of the
    spinal cord.
(C) is active in the resting erect posture.

(D) provides medial rotation in association with extension at the knee joint.
(E) has none of the above properties.

Objectives   10; I; 2; (2.3); 3; (3.6).   10; III; 2; (2.6).
               10; IV; 1; (1.1).   –   –   A

455   Flexion at both hip and knee joints is produced by

(A) the rectus femoris muscle.
(B) the semitendinosus muscle.
(C) the biceps femoris muscle.
(D) the sartorius muscle.
(E) none of the above.

Objectives   10; I; 2; (2.3); 3; (3.6).   –   –   D

456   A useful guide to the level of the knee joint is

(A) the upper border of the tuberosity of the tibia.
(B) the adductor tubercle of the femur.
(C) the upper end of the head of the fibula.
(D) the lower end of the patella.
(E) not provided by the above.

Objectives   10; I; 4; (4.7).   –   –   E

457   The knee joint

(A) is a condylar joint.
(B) is a simple hinge joint.
(C) has a fixed axis around which flexion and extension take place.
(D) has a single axis around which rotation takes place.
(E) permits rotation only in flexion.

Objectives   10; I; 3.   0.38   70   A
                                      0.48   77

458   The posterior cruciate ligament

(A) limits backward slipping of the femur on the
    tibia.
(B) runs forward upwards and laterally from tibia
    to femur.
(C) is completely slack when the leg is fully extended.
(D) limits forward slipping of the femur on the tibia.
(E) limits lateral rotation of the tibia.

Objectives   10; 1; 3; (3.3); (3.4).   0.29   84   D
                                                      0.55   86

459   The tibial collateral ligament of the knee

(A) is cylindrical in shape.
(B) is separated from the medial meniscus by a bursa.
(C) is crossed by the popliteus tendon.
(D) is attached to the medial meniscus.
(E) does not assist in limitation of medial rotation
    of the tibia.

Objectives   10; 1; 3; (3.3), (3.4).   0.44   77   D
                                                      0.47   81

460   The medial meniscus of the knee joint is

(A) more mobile than the lateral meniscus.
(B) not attached to the medial collateral ligament.
(C) less frequently torn than the lateral meniscus.
(D) attached to the tibia but not the femur.
(E) smaller than the lateral meniscus.

Objectives   10; 1; 3.   0.47   72   D
                                                      0.33   51
                                                      0.33   65

461   The medial meniscus of the knee joint is more
      commonly damaged than the lateral because

(A) it is smaller.
(B) the popliteus pulls it out of position.
(C) the femur rotates medially during flexion.

(D) it is firmly attached to the tibial collateral
ligament.
(E) it is pushed posteriorly by the anterior cruciate
ligament.

Objectives   10; 1; 3; (3.3), (3.4).   –   –   D

462   The anterior cruciate ligament of the knee

(A) limits posterior movement of the tibia on the
femur.
(B) limits lateral rotation of the tibia.
(C) resists hyperextension of the knee.
(D) limits overstretching of the posterior cruciate
ligament.
(E) passes upwards, backwards and medially.

Objectives   10; 1; 3; (3.3), (3.4).   –   –   C

463   The anterior cruciate ligament

(A) passes upwards, backwards and laterally.
(B) lies anteromedial to the posterior cruciate
ligament.
(C) limits lateral rotation.
(D) is relaxed in flexion at the knee joint.
(E) is relaxed in extension at the knee joint.

Objectives   10; 1; 3; (3.3), (3.4).   0.44   75   A
                                                0.44   78

A(iii) Ankle and foot (Questions 464–469)

464   The inferior tibio-fibular joint is a

(A) primary cartilaginous joint.
(B) secondary cartilaginous joint.
(C) plane synovial joint.
(D) modified condyloid synovial joint.
(E) fibrous joint.

Objectives   10; 11; 1; (1.6).   0.49   80   E
                                                0.31   66

465   The deltoid ligament

    (A) resists inversion of the foot.
    (B) resists adduction at the ankle joint.
    (C) is attached to the lateral malleolus.
    (D) is attached to the talus.
    (E) is attached to the cuboid.

Objectives   10; II; 1.

0.34	73	D
0.38	85	
0.50	84	

466   On the anterior aspect of the ankle joint the tendon
of the extensor hallucis longus muscle

    (A) is medial to the tibialis anterior muscle.
    (B) is medial to the deep peroneal nerve.
    (C) is lateral to the extensor digitorum longus muscle.
    (D) possesses no synovial sheath.
    (E) is lateral to the extensor hallucis brevis muscle.

Objectives   10; II; 2; (2.8).

0.30	71	B
0.48	77	

467   The ankle joint is least stable when the foot is

    (A) plantar flexed.
    (B) dorsiflexed.
    (C) inverted.
    (D) everted.
    (E) adducted.

Objectives   10; II; 1.

0.38	55	A
0.43	68	

468   Inversion of the foot is performed by

    (A) the peroneus longus and brevis muscles.
    (B) the peroneus longus and tibialis posterior
        muscles.
    (C) the tibialis anterior and tibialis posterior muscles.

(D) the tibialis anterior and peroneus tertius
muscles.
(E) none of the above muscle combinations.

Objectives   10; II; 2; (2.5).   0.46   96   C
                                                                  0.48   89

469   The transverse tarsal joint

(A) is between the talus and calcaneus.
(B) is between the talus and navicular.
(C) comprises the talonavicular and calcaneo-cuboid
joints.
(D) is a purely fibrous joint.
(E) permits dorsiflexion of the foot.

Objectives   10; II; 1; (1.5), (1.7).   0.56   91   C

A(iv) Blood supply and lymphatics (Questions 470–476)

470   The femoral artery

(A) is the continuation of the internal iliac artery.
(B) has muscle cover throughout its course in the
thigh.
(C) gives off its profunda branch in the subsartorial
canal.
(D) lies laterally within the femoral sheath.
(E) lies medial to the saphenous opening.

Objectives   10; III; 1; (1.3).   0.61   87   D
                                                                  0.53   77

471   The femoral canal

(A) is the lateral compartment of the femoral sheath.
(B) is medial to the pubic tubercle.
(C) is medial to the femoral vein.
(D) contains the femoral artery.
(E) contains the femoral nerve.

Objectives   10; III; 1; (1.1).   0.60   94   C
                                                                  0.50   92
                                                                  0.40   87

472 From medial to lateral, the contents of the femoral triangle are

(A) femoral vein, femoral artery, femoral nerve, lymphatics.
(B) lymphatics, femoral vein, femoral artery, femoral nerve.
(C) lymphatics, femoral artery, femoral vein, femoral nerve.
(D) lymphatics, femoral nerve, femoral artery, femoral vein.
(E) femoral nerve, femoral artery, femoral vein, lymphatics.

Objectives   10; III; 1; (1.1), (1.3), (1.6), (1.8).
             10; III; 2; (2.3).   –   –   B

473 Structures which pass through the lesser sciatic foramen include the

(A) obturator internus tendon.
(B) inferior gluteal vessels.
(C) posterior cutaneous nerve of the thigh.
(D) piriformis muscle.
(E) superior gluteal vessels.

Objectives   10; III; 1.   0.42   82   A
                                                      0.54   72

474 A structure which does not lie within the popliteal fossa is the

(A) popliteal artery.
(B) short saphenous vein.
(C) saphenous nerve.
(D) common peroneal nerve.
(E) tibial nerve.

Objectives   10; III; 1; (1.3), (1.6).   0.48   70   C
             10; III; 2; (2.2), (2.3).   0.67   87
                                                      0.57   79

475   The posterior tibial artery is best palpated

    (A) behind the lateral malleolus.
    (B) in the calf.
    (C) at the head of the fibula.
    (D) behind the medial malleolus.
    (E) in front of the ankle joint.

Objectives   10; III; 1; (1.4).
    0.47 65 D
    0.50 94
    0.57 92

476   The dorsalis pedis pulse is normally palpable immediately

    (A) lateral to the tendon of the tibialis anterior muscle.
    (B) medial to the tendon of the extensor hallucis longus muscle.
    (C) lateral to the tendon of the extensor hallucis longus muscle.
    (D) lateral to the tendon of the extensor digitorum longus.
    (E) anterior to the medial malleolus.

Objectives   10; III; 1; (1.4).
    0.32 52 C

A (v) Nerve supply (Questions 477–492)

477   The lumbo-sacral trunk

    (A) is constituted by L5 and S1.
    (B) contributes fibres to the tibial but not the peroneal division of the sciatic nerve.
    (C) contributes fibres to the pudendal nerve.
    (D) contributes fibres to the obturator nerve.
    (E) carries fibres concerned in the movement of inversion.

Objectives   10; II; 2; (2.1), (2.6).
    0.35 76 E

478   Lumbosacral plexus

(A) is made up of ventral rami of all the lumbar and
sacral nerves.
(B) includes the lumbo-sacral trunk made up of
branches from L5 and S1.
(C) gives off the sciatic nerve from S1 to S4.
(D) gives off the femoral nerve from L2 to L4.
(E) gives off the pudendal nerve from S1 to S3.

Objectives   10; III; 2; (2.1).   *0.53*   92   D

479   The femoral nerve

(A) is distributed to skin on the medial side of the
leg.
(B) supplies the adductor magnus muscle.
(C) is distributed to skin on the lateral side of the
thigh.
(D) is derived from the dorsal (primary) rami of
L2, 3 and 4.
(E) is the nerve of the flexors of the knee.

Objectives   10; III; 2; (2.1), (2.5).   *0.73*   75   A
                                                    0.52   66

480   The femoral nerve

(A) arises from the ventral divisions of the ventral
rami of L2, 3 and 4.
(B) supplies the gracilis muscle.
(C) is medial to the femoral artery.
(D) supplies sensation to skin on the medial side
of the leg.
(E) supplies sensation to skin on the outer side of
the thigh.

Objectives   10; III; 2; (2.1), (2.3), (2.5).   −   −   D

481   The obturator nerve

    (A) supplies skin on the lateral side of the thigh.
    (B) carries fibres from S2, 3 and 4.
    (C) traverses the femoral triangle.
    (D) supplies the adductor magnus.
    (E) supplies the obturator internus.

Objectives   10; III; 2; (2.1), (2.3), (2.5).   –   –   D

482   The obturator nerve

    (A) has no cutaneous distribution.
    (B) is the sole nerve supply to adductor magnus.
    (C) does not supply the obturator externus.
    (D) contains fibres from L4.
    (E) supplies the rectus femoris.

Objectives   10; III; 2; (2.1), (2.5).   *0.70*   96   D

483   To avoid the sciatic nerve, an injection into the
    buttock is best given

    (A) into the upper lateral quadrant.
    (B) into the upper medial quadrant.
    (C) into the lower medial quadrant.
    (D) into the lower lateral quadrant.
    (E) at the junction of the four quadrants.

Objectives   10; III; 2.   *0.26*   81   A

484   A muscle not supplied by the medial (tibial)
    division of the sciatic nerve is the

    (A) semimembranosus muscle.
    (B) long head of the biceps femoris muscle.
    (C) adductor magnus muscle.
    (D) semitendinosus muscle.
    (E) short head of biceps femoris muscle.

Objectives   10; III; 2.   *0.28*   64   E

485 The surface markings of the sciatic nerve include the

(A) superior point of trisection of a line from the posterior superior iliac spine to the greater trochanter.
(B) inferior point of trisection of the line described in (A).
(C) superior point of trisection of a line from the posterior superior iliac spine to the ischial tuberosity.
(D) mid-point between the posterior inferior iliac spine and the greater trochanter.
(E) mid-point of a line between the posterior and anterior superior iliac spines.

Objectives   10; III; 2; (2.4).   *0.23*   49   C

486 Division of the sciatic nerve would result in loss of sensation

(A) on all of the thigh, leg and foot.
(B) on the back of the thigh, and calf, and on the sole of the foot.
(C) on all of the leg and foot.
(D) on the lateral side of the calf and most of the foot.
(E) over most of the sole of the foot alone.

Objectives   10; III, 2; (2.5).   *0.57*   72   D
                                                              0.65   75

487 The deep peroneal nerve is the nerve of the

(A) hamstring compartment.
(B) adductor compartment.
(C) muscles of the calf.
(D) anterior compartment of the leg.
(E) peroneal compartment of the leg.

Objectives   10; III; 2.   *0.76*   92   D

488   The deep peroneal nerve supplies

   (A) popliteus.
   (B) plantaris.
   (C) skin on the lateral side of the dorsum of the foot.
   (D) peroneus longus.
   (E) peroneus tertius.

   Objectives   10; III; 2; (2.5).   —   —   E

489   A muscle which usually has a double nerve supply
      is the

   (A) obturator externus muscle.
   (B) adductus magnus muscle.
   (C) semimembranosus muscle.
   (D) pectineus muscle.
   (E) biceps femoris muscle.

   Objectives   10; III; 2.   —   —   B

490   A patient has foot drop and anaesthesia over the
      dorsum of the foot except on the lateral side. The
      lesion is likely to involve the

   (A) tibial and superficial peroneal nerves.
   (B) common peroneal nerve.
   (C) lumbo-sacral trunk.
   (D) the ventral rami of S1 and 2.
   (E) the deep peroneal nerve.

   Objectives   10; III; 2; (2.5), (2.6).   0.47   24   B

491   The spinal nerve supplying the middle of the
      dorsum of the foot is

   (A) L2
   (B) L3
   (C) L5.
   (D) S2.
   (E) S3.

   Objectives   10; III; 2; (2.6).   0.35   97   C

492   The second sacral nerve

    (A) is distributed to skin on the lateral side of the
        sole of the foot.
    (B) is distributed to dorsiflexors of the ankle joint.
    (C) is distributed to invertors of the foot.
    (D) is distributed to extensors of the hip.
    (E) has none of the above properties.

Objectives   10; III; 2; (2.6).   *0.35*   56   E

Type E items (Hubbard and Clemans, 1961) (Questions 493–522)

These items are of the assertion–reason type and there are five possible
responses. The instruction for these items is as follows:

*Each question consists of an assertion and a reason. Responses should be
chosen as follows:*

    *(A) if the assertion and reason are true statements and the reason is a
        correct explanation of the assertion;*
    *(B) if the assertion and reason are true statements but the reason is
        NOT a correct explanation of the assertion;*
    (C) if the assertion is true but the reason is a false statement;
    (D) if the assertion is false but the reason is a true statement;
    (E) if both assertion and reason are false statements.

E(i) Hip (Questions 493–499)

493   The range of extension at the hip joint is small

    BECAUSE

    the buttock stops the backward swing of the thigh
    in extension.

Objectives   10; I; 1; (1.11).   *0.36*   68   C
                                            0.24   81

494   It is structurally advantageous that the iliofemoral
    ligament is the strongest ligament of the hip joint

    BECAUSE

    the hip is most unstable in the extended position.

Objectives   10; I, 1; (1.10).   *0.48*   70   C
                                            0.23   61

495 It is difficult to dislocate the head of the femur
except inferiorly

BECAUSE

the posteriorly located iliofemoral ligament resists
its displacement.

Objectives   10; I; 1; (1.10).   –   –   E

496 During the swing phase of walking (foot off the
ground) the pelvis does not completely sag on
the swing side

BECAUSE

the abductors at the hip on the opposite side contract
strongly during the swing phase.

Objectives   10; IV; 1; (1.3).   0.39   91   A
                                                          0.34   85
                                                          0.50   87

497 Gluteus maximus paralysis results in a marked limp

BECAUSE

this muscle is a powerful extensor of the hip joint.

Objectives   10; IV; 1.   0.64   86   D

498 A lesion of the right superior gluteal nerve results
in a marked pelvic tilt on the right in the swing
phase

BECAUSE

this nerve supplies the gluteus medius and minimus,
important abductors at the hip joint.

Objectives   10; IV; 1; (1.3).   0.61   66   D

499   The hamstring muscles are used only in strong
      extension of the thigh at the hip, but not as extensors
      in normal walking

      BECAUSE

      the gluteus maximus muscle is used for extension at
      the hip joint in normal walking.

      Objectives   10; IV; 1; (1.3).   *0.60*   73   E
                                                             0.26   82

E(ii) Knee (Questions 500–508)

500   The quadriceps femoris muscle shows little or no
      activity in the resting erect posture

      BECAUSE

      the line of gravity is in front of the centre of the
      knee joint.

      Objectives   10; IV; 1; (1.1).   *0.30*   67   A
                                                             0.45   88
                                                             0.39   50

501   The adductor magnus muscle does not act on the
      knee joint

      BECAUSE

      it has no attachment to either tibia or fibula.

      Objectives   10; I; 2.   *0.33*   63   A

502   The semimembranosus and the short head of the
      biceps femoris act as flexors at the knee joint

      BECAUSE

      they share a common proximal attachment to
      the hip bone and are inserted distal to the knee
      joint.

      Objectives   10; I; 3; (3.6).   *0.47*   73   C

503   Accumulation of fluid in the knee joint may cause
a visible swelling above the knee

BECAUSE

the suprapatellar pouch (bursa) communicates with
the knee joint.

Objectives   10; l; 3; (3.3).   0.27   88   A
                                                                        0.40   93

504   The anterior cruciate ligament becomes tight as the
leg is extended

BECAUSE

the tibia is displaced anteriorly on the femur during
extension at the knee.

Objectives   10; l; 3; (3.4).   −   −   A

505   The popliteus muscle is considered to be important
in unlocking the knee joint

BECAUSE

acting from a fixed tibia it can rotate the femur
medially.

Objectives   10; l; 3.   0.51   79   C

506   Lateral rotation of the tibia on the femur occurs as
the leg is fully extended (when the foot is free)

BECAUSE

contraction of the biceps femoris produces lateral
rotation at the knee joint.

Objectives   10; l; 3.   0.38   60   B
                                                                        0.41   47
                                                                        0.58   59

507   The tibia rotates on the femur at the commencement
      of flexion at the knee joint when the foot is off the
      ground

      BECAUSE

      it is necessary to unlock the rotation which occurs
      in extension.

      Objectives   10; I; 3.   *0.49*   76   **A**

508   Rotary movements at the knee joint are freer in
      flexion

      BECAUSE

      in this position the fibular collateral ligament and
      the posterior part of the tibial collateral ligament
      are relaxed.

      Objectives   10; I; 3.   −   −   **A**

E(iii) Foot and ankle (Questions 509–512)

509   The peroneal muscles help to maintain the plantigrade
      position of the foot

      BECAUSE

      they act as evertors at the ankle joint.

      Objectives   10; II; 1; (1.7).   *0.43*   32   **C**

510   Division of the tendo calcaneus is not a serious
      injury

      BECAUSE

      the long flexors of the toes are powerful plantar
      flexors of the ankle.

      Objectives   10; II; 2; (2.4).   −   −   **E**

511   The tibialis posterior muscle is an invertor of the
foot

BECAUSE

it acts on the subtaloid joints medial to their axis
of movement.

Objectives   10; II; 2.   *0.28*   74   A

512   The movement of inversion takes place mainly at
the talocalcaneal and talocalcaneonavicular joints

BECAUSE

the talus is virtually immobile in this movement
and the calcaneus carries the rest of the foot
with it.

Objectives   10; II; 1; (1.7).   *0.48*   58   A
                                                            0.43   66

E(iv) Blood supply and lymphatics (Questions 513–515)

513   The whole lymphatic drainage of the lower limb
traverses the external iliac nodes

BECAUSE

these nodes receive efferents from both the
superficial and deep inguinal lymph nodes.

Objectives   10; III; 1; (1.8).   –   –   D

514   The superficial inguinal lymph nodes do not collect
lymph from the buttock

BECAUSE

they collect lymph only from below the inguinal
ligament.

Objectives   10; III; 1; (1.8).   *0.30*   74   E
                                                            0.39   74

515 Perforating veins are important channels carrying
blood from deep to superficial venous
circulations in the lower limb

BECAUSE

in the lower limb most of the venous return takes
place through superficial veins.

Objectives   10; III; 1; (1.7).   0.51   55   E
                                                   0.44   62
                                                   0.58   76

E(v) Nerve supply (Questions 516–522)

516 The femoral nerve remains outside the femoral sheath

BECAUSE

it arises behind the fascia covering the iliacus muscle.

Objectives   10; III; 2.   0.33   61   A

517 A lesion of the femoral nerve in the lower third of
the thigh makes active extension of the knee
impossible

BECAUSE

the femoral nerve supplies the quadratus femoris
muscle.

Objectives   10; III; 2.   –   –   E

518 The ischial part of the adductor magnus is innervated
by the tibial component of the sciatic nerve

BECAUSE

it is derived from the hamstring group of muscles
developmentally.

Objectives   10; III; 2.   0.40   84   A

519   Fractures of the neck of the fibula could be
      expected to result in inability to dorsiflex the foot

      BECAUSE

      the common peroneal nerve may be injured where
      it lies medial to the neck of the fibula.

      Objectives   10; III; 2; (2.2), (2.5).   *0.45*   33   C
                                                    0.59   73
                                                    0.59   68

520   Damage to the common peroneal nerve may result in
      inability to dorsiflex the foot

      BECAUSE

      the common peroneal nerve supplies all the muscles
      in the flexor compartment of the calf.

      Objectives   10; III; 2; (2.5).   *0.39*   72   C
                                                    0.55   81
                                                    0.45   73

521   Division of the tibial nerve results in paralysis of
      the soleus

      BECAUSE

      the tibial nerve is distributed to the plantar flexors
      of the ankle which include this muscle.

      Objectives   10; III; 2; (2.5).   *0.29*   85   A
                                                    0.50   95

522   Division of the tibial nerve results in 'foot drop'
      BECAUSE

      the tibial nerve supplies the gastrocnemius and
      soleus.

      Objectives   10; III; 2; (2.5).   *0.58*   85   D

Type K items (Hubbard and Clemans, 1961) (Questions 523–570)

These items are in the form of incomplete statements and call for recognition of one or more correct responses which are grouped in various ways. The instruction for these items is as follows:

For each of the incomplete statements below, one or more of the completions given is correct. Responses should be chosen as follows:
(A) if only completions (i), (ii) and (iii) are correct,
(B) if only completions (i) and (iii) are correct,
(C) if only completions (ii) and (iv) are correct,
(D) if only completion (iv) is correct,
(E) if all completions are correct.

K(i) Hip (Questions 523–536)

523 At the hip joint

(i) the position of maximum stability is in extension, abduction and medial rotation.
(ii) flexion may be limited by tension in the hamstring muscles.
(iii) extension is associated with extension of the lumbar spine.
(iv) the female articular surface is cup-shaped and fully lined by hyaline cartilage.

Objectives 10; I; 1.   0.20 43   A
                             0.49 51

524 The neck of the femur

(i) is bounded posteriorly by the intertrochanteric crest.
(ii) is only partially intracapsular.
(iii) carries the femoral head in a position of anteversion.
(iv) is lined by synovial membrane more extensively anteriorly than posteriorly.

Objectives 10; I; 1; (1.7), (1.8), (1.9).   0.30 48   E
                                             0.32 35

525 The head of the femur

    (i)   has an articular surface of more than half a sphere

    (ii) is wholly intracapsular.

    (iii) transmits the body weight to the lower limb.

    (iv) has a ligament attached to it.

Objectives   10; I; 1; (1.8).   *0.44*   88   E

526 The hip joint owes its stability to

    (i)   the diameter of the femoral head being greater than that of the rim of the acetabular labrum.

    (ii) the muscles surrounding the joint being fused with the capsule.

    (iii) the tautness of the iliofemoral ligament in extension.

    (iv) the strength of the ligament of the head of the femur.

Objectives   10; I; 1.   *0.42*   58   B

527 In radiographs of the hip joint in the anatomical position

    (i)   the tip of the greater trochanter is in the same horizontal plane as the middle of the head of the femur.

    (ii) the tip of the greater trochanter is in the same horizontal plane as the pubic tubercle.

    (iii) the inferior surfaces of the femoral neck and head form a curve continuous with the inferior border of the superior ramus of the pubis (Shenton's line).

    (iv) the lesser trochanter is only partly visible.

Objectives   10; I; 1; (1.12).   *0.36*   46   E
                                                                  0.40   51

528   Landmarks easily felt in clinical examination
       of the hip joint include the

       (i)   greater trochanter of the femur.
       (ii) ischial tuberosity.
       (iii) anterior superior iliac spine.
       (iv) lesser trochanter of femur.

       Objectives   10; I; 1; (4.1).   0.29   76   A
                                                   0.35   95
                                                   0.27   95

529   The gluteus maximus muscle

       (i)   is a principal abductor at the hip joint.
       (ii) acts on the knee joint via the iliotibial tract.
       (iii) contracts strongly during normal walking on flat
             ground.
       (iv) is an extensor of the trunk on the lower limb.

       Objectives   10; I; 2.   0.43   76   C
                                                   0.20   56

530   The gluteus maximus muscle is

       (i)   chiefly used during ordinary walking movements.
       (ii) used during the action of sitting.
       (iii) a flexor at the hip joint.
       (iv) important during running and climbing.

       Objectives   10; I; 2.   0.36   64   C
                                                   0.23   79

531   The semimembranous muscle

       (i)   is a flexor at the knee joint.
       (ii) forms the upper medial boundary of the popliteal
             fossa.
       (iii) is innervated by the tibial component of the
             sciatic nerve.
       (iv) is a medial rotator of the tibia.

       Objectives   10; I; 2.   0.33   88   E
                                                   0.35   72

532 Adductors of the hip include the

(i)   quadratus femoris muscle.
(ii) gluteus medius muscle.
(iii) pectineus muscle.
(iv) sartorius muscle.

Objectives   10; I; 2; (2.3).   —   —   B

533 The adductor magnus muscle

(i)   is a weak adductor of the thigh.
(ii) is a weak extensor at the hip.
(iii) is supplied by the obturator nerve and the
      common peroneal component of the sciatic
      nerve.
(iv) is supplied by the obturator nerve and the tibial
      component of the sciatic nerve.

Objectives   10; I; 2.   *0.52* 81   C

534 The lateral rotators at the hip include the

(i)   piriformis muscle.
(ii) quadratus femoris muscle.
(iii) obturator internus muscle.
(iv) obturator externus muscle.

Objectives   10; I; 2; (2.3).   *0.73* 85   E

535 With reference to the swinging lower limb in normal
     walking along level ground, there is

(i)   flexion at the hip joint.
(ii) medial rotation of the tibia on the femur.
(iii) lateral rotation at the hip joint.
(iv) a shift of the centre of gravity of the body
      to that side.

Objectives   10; IV; 1; (1.3).   *0.17* 39   B

536 In the anatomical position the line of gravity passes

(i)   through the transverse axis of the hip joint.
(ii) behind the transverse axis of the hip joint.
(iii) behind the transverse axis of the knee joint.
(iv) in front of the transverse axis of the knee joint.

Objectives   10; IV; 1; (1.1).   *0.43*   93   C

K(ii) Knee region (Questions 537–547)

537 With regard to the quadriceps femoris muscle

(i)   the rectus femoris component functions as a
      flexor at the hip.
(ii) the main functional attachment at the lower end
      is into the tibia via the ligamentum patellae.
(iii) nerve supply comes from lumbar spinal nerves.
(iv) the lowermost muscle fibres of vastus medialis
      muscle are almost horizontally arranged.

Objectives   10; I; 3; (3.5).   *0.61*   85   E
                                                   0.36   91

538 The capsule of the knee joint

(i)   is reinforced by expansions from surrounding
      muscles and tendons.
(ii) is penetrated by the cruciate ligaments.
(iii) is attached to the menisci at their peripheral
      margins.
(iv) is applied throughout to the underlying synovial
      membrane.

Objectives   10; I; 3.   *0.28*   63   B
                                                   0.32   59

539 The patella

(i)   may be considered as a sesamoid bone in the
      tendon of the quadriceps femoris.

(ii) is stabilized from lateral displacement by the greater prominence of the lateral femoral condyle.
(iii) gives attachment to fibres of the vastus medialis muscle.
(iv) has a larger medial articular surface as compared with lateral articular surface.

Objectives   10; 1; 3.   0.40   45   A
                                                0.40   40

540 The hamstring muscles

(i)   are all (except short head of biceps) innervated by the peroneal component of the sciatic nerve.
(ii) extend the thigh at the hip joint.
(iii) may extend the thigh on the leg at the knee joint if the leg is fixed.
(iv) often limit extension at the knee joint if the thigh is flexed.

Objectives   10; 1; 2; (2.3); 3; (3.6).   0.65   92   C

541 The medial meniscus

(i)   is more fixed than the lateral meniscus.
(ii) accommodates to the changing profile of the medial femoral condyle in flexion and extension.
(iii) functions as a shock absorber.
(iv) is smaller and more circular than the lateral meniscus.

Objectives   10; 1; 3; (3.3), (3.4).   0.63   62   A
                                                0.46   74

542 The suprapatellar bursa usually

(i)   lies superficial to the quadriceps femoris tendon.
(ii) extends about 15 cm above the patella.

(iii) is the site of fluid accumulation in 'housemaid's
      knee'.
(iv) communicates with the knee joint.

Objectives   10; I; 3; (3.3).   0.24   48   D
                                                       0.45   68

543   A normal subject stands upright at ease. Significant
      postural activity occurs in the

      (i)   iliopsoas muscle.
      (ii) quadriceps femoris muscle.
      (iii) triceps surae muscle.
      (iv) muscles of the sole of the foot.

      Objectives   10; IV; 1; (1.1).   0.41   76   B

544. The quadriceps femoris muscle is

      (i)   supplied by the femoral and obturator nerves.
      (ii) important for walking downstairs.
      (iii) the main flexor at the hip joint.
      (iv) necessary for the knee jerk reflex.

      Objectives   10; I; 3.   0.52   88   C
                                                       0.47   80

545   The iliotibial tract

      (i)   is a lateral thickening in the fascia lata.
      (ii) receives part of the gluteus medius muscle.
      (iii) encloses the tensor fasciae latae muscle.
      (iv) does not extend distal to the knee joint.

      Objectives   10; I; 3.   0.33   78   B
                                                       0.51   77

546   Muscles which flex the leg at the knee joint include
      the

      (i)   popliteus muscle.
      (ii) semimembranosus muscle.

(iii) gastrocnemius muscle.
(iv) biceps femoris muscle.

Objectives   10; I; 3; (3.6).

0.50	68	E
0.40	89	
0.36	82	

547 Factors preventing lateral dislocation of the patella
     include the

(i)   more prominent articular area for the patella on
      the lateral femoral condyle.
(ii) angle between femur and tibia.
(iii) lowest fibres of vastus medialis muscle.
(iv) direction of pull of quadriceps femoris muscle.

Objectives   10; I; 3.   —   —   B

K(iii) Ankle and foot (Questions 548–555)

548 Attached to the tendo calcaneus are the

(i)   soleus muscle.
(ii) medial head of gastrocnemius muscle.
(iii) lateral head of gastrocnemius muscle.
(iv) tibialis posterior tendon.

Objectives   10; II; 2; (2.2).

0.53	93	A
0.85	99	

549 The tibialis posterior muscle

(i)   has a tendon with no synovial sheath.
(ii) is an invertor of the foot.
(iii) has a tendon which grooves the back of the lateral
      malleolus.
(iv) is inserted chiefly distal to the subtaloid joint.

Objectives   10; II; 2; (2.2), (2.5).   —   —   C

550 Muscles in the posterior compartment of the leg

    (i)   include flexors at the knee joint.
    (ii) are all innervated through the tibial nerve.
    (iii) include invertors of the foot.
    (iv) are inactive in maintenance of the resting erect
         posture.

Objectives   10; II; 2; (2.2), (2.5).   10; III; 2; (2.5).   *0.41* 84   A
               10; IV; 1; (1.1).   *0.35* 64

551 The medial (deltoid) ligament is attached to the

    (i)   medial malleolus.
    (ii) sustentaculum tali.
    (iii) 'spring' ligament.
    (iv) tuberosity of the navicular.

Objectives   10; II; 1.   *0.45* 73   E

552 The medial (deltoid) ligament of the ankle joint

    (i)   resists displacement of the foot backwards.
    (ii) resists displacement of the foot forwards.
    (iii) limits plantar flexion of the foot.
    (iv) limits eversion of the foot.

Objectives   10; II; 1.   *0.37* 42   E
                                          0.32 51

553 The sustentaculum tali

    (i)   gives partial attachment to the spring ligament.
    (ii) is a protection from the calcaneus.
    (iii) supports the head of the talus.
    (iv) gives partial attachment to the deltoid ligament.

Objectives   10; II; 1.   *0.43* 27   E
                                          0.30 63
                                          0.34 61

554   Bones normally articulating with the navicular by
      synovial joints include the

      (i) talus.
      (ii) intermediate cuneiform.
      (iii) lateral cuneiform.
      (iv) calcaneus.

      Objectives   10; II; 1; (1.5).   —   —   A

555   Lying anterior to the ankle joint are

      (i) the tibialis anterior tendon.
      (ii) the extensor digitorum longus tendon.
      (iii) the anterior tibial artery.
      (iv) branches of the superficial peroneal nerve.

      Objectives   10; II; 2.   0.32   44   E
                                                 0.31   50
                                                 0.20   53

K(iv) Blood supply and lymphatics (Questions 556–560)

556   The femoral artery

      (i)   continues as the popliteal artery.
      (ii) gives off the profunda femoris artery.
      (iii) is continuous with the external iliac artery.
      (iv) accompanies part of the great saphenous vein.

      Objectives   10; III; 1; (1.3).   0.49   87   A

557   The posterior tibial artery

      (i)   divides into medial and lateral plantar arteries.
      (ii) continues into the foot as the dorsalis pedis artery.
      (iii) is accompanied by the tibial nerve.
      (iv) is accompanied by the deep peroneal nerve.

      Objectives   10; III; 1; (1.3).   0.71   96   B

558   A femoral hernia

(i)   passes through the femoral canal.
(ii) forms a lump below and lateral to the pubic
      tubercle.
(iii) passes through the saphenous opening.
(iv) is closely related to the femoral nerve.

Objectives   10; III; 1; (1.2).   *0.46*   82   A

559   The great saphenous vein

(i)   has no valves in its course in the leg.
(ii) passes through the saphenous opening to join
      the femoral vein.
(iii) is connected with deep veins by valveless
      channels.
(iv) lies in front of the medial malleolus.

Objectives   10; III; 1; (1.6), (1.7).   *0.42*   94   C
                                                   0.61   91

560   The superficial inguinal nodes receive lymph from
      the subcutaneous tissues of the

(i)   lower abdominal wall.
(ii) perineum.
(iii) toes.
(iv) gluteal region.

Objectives   10; III; 1; (1.8).   *0.39*   54   E

K(v) Nerve supply (Questions 561–570)

561   The inferior gluteal nerve supplies the

(i)   gluteus minimus muscle.
(ii) tensor fascia latae muscle.
(iii) gluteus medius muscle.
(iv) gluteus maximus muscle.

Objectives   10; III; 2.   *0.32*   73   D
                                                   0.51   79

562   A patient has suffered fracture of the neck of the
      fibula. Involvement of the common peroneal nerve
      in the injury would be indicated by

      (i)   inability to dorsiflex the foot.
      (ii) inability to plantar flex the foot.
      (iii) anaesthesia of the dorsum of the foot.
      (iv) anaesthesia of the sole.

      Objectives   10; III; 2; (2.5).   *0.24*   94   B

563   Loss of sensation of the skin on the medial aspect of
      the leg may indicate injury to

      (i)   spinal nerves L3 and L4.
      (ii) the common peroneal nerve.
      (iii) the femoral nerve.
      (iv) the deep peroneal nerve.

      Objectives   10; III; 2.   *0.45*   63   B
                                                              0.37   89

564   If the common peroneal nerve is divided

      (i)   eversion of the foot is substantially lost.
      (ii) cutaneous sensation from the sole of the foot is
            lost.
      (iii) dorsiflexion of the foot is lost.
      (iv) there is no loss of cutaneous sensation.

      Objectives   10; III; 2; (2.5).   *0.65*   87   B
                                                              0.45   86
                                                              0.37   69

565   The common peroneal nerve

      (i)   ends by dividing into deep and superficial
            peroneal nerves.
      (ii) may be injured lateral to the head of the fibula.

(iii) forms a discrete component of the sciatic nerve.
(iv) arises from the ventral divisions of the ventral
      rami of L4, L5, S1, S2.

Objectives   10; III; 2; (2.2).   —   —   B

566   The sciatic nerve supplies the

(i)   biceps femoris muscle.
(ii) semitendinosus muscle.
(iii) semimembranosus muscle.
(iv) adductor magnus muscle.

Objectives   10; III; 2; (2.5).   —   —   E

567   The obturator nerve supplies the

(i)   adductor longus muscle.
(ii) obturator externus muscle.
(iii) adductor magnus muscle.
(iv) tensor fasciae latae muscle.

Objectives   10; III; 2; (2.5).   0.22   64   A
                                                  0.51   85

568   Muscles supplied by the femoral nerve include the

(i)   rectus femoris muscle.
(ii) vastus medialis muscle.
(ii) pectineus muscle.
(iv) sartorius muscle.

Objectives   10; III; 2; (2.5).   0.58   88   E
                                                  0.98   97

569   The femoral nerve

(i)   supplies skin on the lateral side of the leg.
(ii) enters the thigh through the obturator foramen.

(iii) contains fibres from the dorsal rami of L2, L3, and L4.

(iv) lies lateral to the femoral vessels.

Objectives   10; III; 2; (2.1), (2.3), (2.5).   —   —   D

570 Muscles supplied by the superficial peroneal nerve include the

(i)   peroneus longus muscle.
(ii) peroneus tertius muscle.
(iii) peroneus brevis muscle.
(iv) flexor hallucis longus muscle.

Objectives   10; III; 2.   *0.52*   83   B
                              0.42   94

Chapter 6

The Cardiovascular and Respiratory Systems

Type A items (Hubbard and Clemans, 1961) (Questions 571–593)

These involve choosing one correct answer from five available choices. The instruction for these items is as follows:
Each of the incomplete statements below is followed by five suggested answers or completions. Select the one which is best in each case.

A(i) External and internal features of the heart (Questions 571–577)

571 The base of the heart faces

(A) anteriorly and to the right.
(B) superiorly.
(C) posteriorly and to the right.
(D) inferiorly.
(E) in none of the above directions.

Objectives   11; I; 1; (1.2).   *0.39*   64   C
                                        0.26   70
                                        0.31   66

572 In the adult the right atrium

(A) does not extend beyond the right margin of the sternum.

165

(B) is partly separated from the right ventricle
by the interventricular septum.
(C) has a bicuspid valve leading from it into the
right ventricle.
(D) extends about 5 cm from the right margin of
the sternum.
(E) has the coronary sinus draining into it.

Objectives   11; I; 1; (1.1), (1.2), (1.4), (1.10).   *0.75   95*   E
                                                          0.54   98

573   The fossa ovalis is in

(A) the posterior wall of the left atrium.
(B) the left ventricle on the interventricular septum.
(C) the right ventricle on the interventricular septum.
(D) the anterior wall of the right atrium.
(E) none of the above.

Objectives   11; I; 1; (1.1), (1.4—1.8).   *0.23   84*   E
                                              0.57   71
                                              0.35   90

574   The right ventricle

(A) forms most of the right border of the heart.
(B) has thicker walls than the left ventricle.
(C) lies anterior in plane to the right atrium.
(D) is ridged throughout by trabeculae carneae.
(E) receives the opening of the coronary sinus.

Objectives   11; I; 1; (1.1), (1.2), (1.5).   *0.52   74*   C
                                                 0.40   50
                                                 0.60   84

575   The right and left ventricles show differences in
all of the following, except

(A) the thickness of their walls.
(B) their relationships to the ribs.

    (C) the sizes of their respective atrioventricular
        orifices.
    (D) the arrangement and number of their papillary
        muscles.
    (E) their volume outputs.

Objectives   1; I; 1; (1.2), (1.5), (1.6).   —   —   E

576   The interventricular septum

    (A) lies in a sagittal plane.
    (B) is muscular throughout.
    (C) contains the atrioventricular node.
    (D) is convex anteriorly and to the right.
    (E) has a septal cusp of the pulmonary valve
        attached to it.

Objectives   1; I; 1; (1.8).   11; I; 2; (2.8).   *0.52*   74   D

577   The pulmonary valve

    (A) lies posterior to the aortic valve.
    (B) lies between the right ventricle and the pulmonary
        veins.
    (C) transmits highly oxygenated blood.
    (D) has three cusps.
    (E) has none of the above properties.

Objectives   1; I; 1; (1.3), (1.5), (1.12).   *0.37*   60   D
                                                           0.32   88

A(ii) Blood and nerve supply of the heart, pericardium, lymphatics (Questions 578–586)

578   Veins draining the heart empty

    (A) directly into the atria alone.
    (B) into the coronary sinus and directly into the
        right ventricle.
    (C) via the anterior cardiac vein into the left atrium.

(D) into the coronary sinus and directly into the chambers of the heart.

(E) into the coronary sinus alone.

Objectives   1; 1; 2; (2.4).   *0.39*   90   D
                                                 0.46   79
                                                 0.50   92

579   The cardiac plexus

(A) controls heart rhythm.

(B) contains visceral efferent fibres only.

(C) is the site of origin of postganglionic sympathetic fibres.

(D) is concerned with reflex control of heart rate.

(E) receives postganglionic parasympathetic vagal fibres.

Objectives   1; 1; 2; (2.6), (2.7).   *0.42*   68   D
                                                 0.44   69

580   The sinu-atrial node is situated

(A) on the right of the opening of the inferior vena cava.

(B) within the inter-atrial septum.

(C) at the opening of the coronary sinus.

(D) at the opening of the upper right pulmonary vein.

(E) in front of the opening of the superior vena cava.

Objectives   1; 1; 2; (2.8).   *0.23*   89   E

581   In relation to an electrocardiograph

(A) the PR interval is largely produced by delay in junctional tissue around the A–V node.

(B) the T wave represents atrial depolarization.

(C) the major part of the interval between successive QRS complexes is occupied by systole.

(D) the P wave represents ventricular depolarization.
(E) none of the above statements is correct.

Objectives   11; I; 2; (2.10).   *0.31*   73   A
                                                             0.46   67

582 The fibrous pericardium

(A) has visceral and parietal layers.
(B) has no external attachments.
(C) encloses part only of the superior vena cava.
(D) is related inferiorly to the diaphragmatic pleura.
(E) has none of the above properties.

Objectives   11; I; 1; (1.3).   11; I; 2; (2.11), (2.12).   *0.35*   74   C
                                                               0.35   61

583 The fibrous pericardium

(A) encloses the arch of the aorta.
(B) is completely separated from the sternum by the
    pleura.
(C) is fused with the central tendon of the diaphragm.
(D) is related posteriorly to the 3rd thoracic vertebra.
(E) in innervated mainly by the vagus nerve.

Objectives   11; I; 1; (1.3).   11; I; 2; (2.11), (2.12).   *0.45*   93   C
                                                               0.39   95

584 Lymphatic vessels

(A) are of uniform calibre.
(B) always accompany blood vessels.
(C) contain numerous valves.
(D) are present in all tissues.
(E) have none of the above properties.

Objectives   1; II; 1; (1.1), (1.3).   −   −   C

585 Flow of lymph is

(A) determined by the heart rate.
(B) dependent on gravity.
(C) obstructed by lymph nodes.
(D) at a uniform rate throughout the body.
(E) unidirectional within a particular lymphatic vessel.

Objectives   11; II; 1; (1.1–1.3).   *0.19*   85   E

586 The common termination of the thoracic duct is into the

(A) superior vena cava.
(B) junction of left subclavian and left internal jugular veins.
(C) right brachiocephalic vein.
(D) termination of vena azygos.
(E) right subclavian vein.

Objectives   1; II; 1; (1.4).   *0.42*   87   B

A(ii) Respiratory system (Questions 587–593)

587 In the living subject the trachea bifurcates

(A) at the level of the lung roots.
(B) at the level of the manubriosternal joint.
(C) in the mid-line.
(D) into bronchi of equal size and length.
(E) to the right of the mid-line.

Objectives   11; III; 1; (1.1).   *0.31*   76   E

588 The trachea is

(A) in the mid-line throughout its course.
(B) an anterior relation to the left brachiocephalic vein.
(C) an anterior relation of the brachiocephalic trunk.

(D) a posterior relation of the aortic arch.
(E) in none of the above positions.

Objectives   1; III; 1; (1.1).   *0.37*   74   D

589   The right lung

(A) is grooved by the arch of the aorta.
(B) has three bronchopulmonary segments in its
    middle lobe.
(C) is grooved by the azygos vein.
(D) possesses a lingula.
(E) has none of the above properties.

Objectives   11; I; 1; (1.3).   11; III; 1; (1.2), (1.4).   *0.60*   95   C

590   The left lung

(A) has upper, middle and lower lobes.
(B) possesses a lingula.
(C) is larger than the right lung.
(D) is grooved by the trachea.
(E) has none of the above properties.

Objectives   11; III; 1; (1.1), (1.2).   *0.36*   77   B
                                                  0.46   93

591   A bronchopulmonary segment

(A) is supplied by a 1st division bronchus.
(B) is ventilated by a single bronchiole.
(C) is separated from adjacent bronchopulmonary
    segments by connective tissue septa.
(D) is drained mainly by the bronchial venous system.
(E) can be distinguished on a straight antero-posterior
    radiograph of the thorax.

Objectives   11; III; 1; (1.1), (1.4–1.6).   *0.33*   37   C

592   The pleural cavities normally

(A) communicate via the transverse sinus.
(B) communicate with the peritoneal cavity.
(C) extend below the 12th rib.
(D) contain air at less than atmospheric pressure.
(E) lie entirely superior to the peritoneal cavity.

Objectives   11; I; 2; (2.14).   11; III; 2; (2,1), (2.5).   *0.59*   85   C

593   A horizontal stab 4 inches deep is made in the 7th
right intercostal space in the mid-clavicular line.
Structures (deep into skin and muscle) are then
pierced in the order of

(A) parietal pleura, visceral pleura and lung.
(B) two layers of parietal pleura, visceral pleura and
lung.
(C) parietal pleura, visceral pleura, lung, visceral
pleura, diaphragmatic pleura and diaphragm.
(D) two layers of perietal pleura and diaphragm.
(E) none of the above.

Objectives   11; III; 1; (1.2).
11; III; 2; (2.1), (2.3), (2.5).   —   —   C

Type E items (Hubbard and Clemans, 1961) (Questions 594—603)

These items are of the assertion—reason type and there are five possible
responses. The instruction for these items is as follows:
*Each question consists of an assertion and a reason. Responses should be
chosen as follows:*
*(A) if the assertion and reason are true statements and the reason is a
correct explanation of the assertion;*
*(B) if the assertion and reason are true statements but the reason is
NOT a correct explanation of the assertion;*
(C) if the assertion is true but the reason is a false statement;
(D) if the assertion is false but the reason is a true statement;
(E) if both assertion and reason are false statements.

E(i) External and internal features of the heart (Questions 594—598)

594   The best place for hearing sounds transmitted
from the aortic valve is in the second right intercostal
space close to the sternum

BECAUSE

the aortic orifice is located directly deep to this
situation.

Objectives   11; I; 1.   *0.43*   93   C

595   The first heart sound should be heard with maximum
clarity in the second left intercostal space

BECAUSE

it is produced by closure of the atrioventricular valves.

Objectives   11; I; 1; (1.10—1.12).   *0.56*   86   D

596   It is necessary to use heavy percussion to demonstrate
the left border of the area of cardiac dullness

BECAUSE

except at the cardiac notch the left side of the
sternocostal surface of the heart is overlapped by
lung tissue.

Objectives   11; I; 1; (1.2), (1.10).   11; III; 1; (1.2).   *0.21*   68   A
11; III; 2; (2.5), (2.7).   *0.36*   84

597   The walls of the cardiac ventricles are the same
thickness

BECAUSE

both sides of the heart pump the same stroke volume
of blood.

Objectives   11; I; 1; (1.5), (1.6), (1.12).   *0.39*   84   D

598 The second heart sound should be heard with maximum clarity at the apex beat

BECAUSE

it is produced by closure of the aortic and pulmonary valves.

Objectives   11; I; 1; (1.10–1.12).   *0.36*   52   D

E(ii) Blood and nerve supply of heart, pericardium, lymphatics (Questions 599–600)

599 The P wave of an ECG is correlated with atrial contraction

BECAUSE

it is generated by depolarization of atrial muscle.

Objectives   11; I; 2; (2.10).   *0.34*   88   A
                                                  0.49   67

600 The pericardial cavity can be aspirated without transgression of the pleura

BECAUSE

the pericardium is directly related to the left 5th intercostal space adjacent to the sternum.

Objectives   11; I; 2; (2.11), (2.13).   *0.34*   59   A
             11; III; 2; (2.1), (2.5).   *0.46*   46

E(iii) Respiratory system (Questions 601–603)

601 Breath sounds have a transmitted 'tubular' character at the right apex

BECAUSE

no major vessels intervene between the mediastinal aspect of the right lung and most of the intrathoracic part of the trachea.

Objectives   11; III; 1; (1.1), (1.2).
             11; III; 2; (2.5), (2.6), (2.8).   –   –   A

602   Irritation of the pleura, which lies on the peripheral
      part of the diaphragm, results in pain referred to the
      ipsilateral shoulder

      BECAUSE

      the diaphragmatic pleura is mainly supplied by the
      phrenic nerve.

      Objectives   11; III; 2; (2.1), (2.4).   –   –   D

603   Diaphragmatic pleurisy may give rise to pain in the
      ipsilateral shoulder

      BECAUSE

      the diaphragmatic parietal pleura is mainly supplied
      by the phrenic nerve (C3, 4, 5).

      Objectives   11; III; 2; (2.1), (2.4).   *0.37*   81   A

Type K items (Hubbard and Clemans, 1961) (Questions 604–640)

These items are in the form of incomplete statements and call for recognition
of one or more correct responses which are grouped in various ways. The
instruction for these items is as follows:
      For each of the incomplete statements below, one or more of the
      completions given is correct. Responses should be chosen as follows:
      (A) if only completions (i), (ii) and (iii) are correct,
      (B) if only completions (i) and (iii) are correct,
      (C) if only completions (ii) and (iv) are correct,
      (D) if only completion (iv) is correct,
      (E) if all completions are correct.

K(i) External and internal features of the heart (Questions 604–618)

604   The arch of the aorta

      (i)   begins opposite the left third costal cartilage at its
            junction with the sternum.

(ii) extends superiorly to the suprasternal notch.
(iii) gives off the coronary arteries.
(iv) gives off the left subclavian artery.

Objectives   11; I; 1; (1.3).   11; I; 2; (2.1).   *0.32*   52   D
                                                        0.47   66

605   The arch of the aorta

(i)   is crossed anteriorly and to the left by the left
      phrenic nerve.
(ii) lies almost in a sagittal plane.
(iii) has the left brachiocephalic vein above it.
(iv) usually causes an impression on the left side of
      the oesophagus.

Objectives   11; I; 1.   —   —   E

606   The thoracic aorta

(i)   has the thoracic duct anteriorly.
(ii) usually has the azygos vein to its left.
(iii) enters the abdomen to the left of the mid-line.
(iv) leaves the thorax at the level of the 12th thoracic
      vertebra.

Objectives   11; I; 1; (1.3).   11; II; 1; (1.4).   —   —   D

607   The thoracic aorta

(i)   lies directly anterior to the vertebral column
      throughout its course.
(ii) is crossed by the oesophagus.
(iii) pierces the central tendon of the diaphragm.
(iv) gives off nine pairs of posterior intercostal
      arteries.

Objectives   11; I; 1.   *0.52*   64   C

608   The superior vena cava

    (i)   is entirely extrapericardial.
    (ii) lies anterior to the transverse sinus of the
        pericardium.
    (iii) has the right vagus nerve on its lateral surface.
    (iv) receives the azygos vein.

Objectives   11; I; 1.   11; I; 2.

0.27	33	D
0.29	59	
0.45	78	

609   The superior vena cava

    (i)   has no valves.
    (ii) is formed by union of the brachiocephalic veins.
    (iii) lies partly within the pericardium.
    (iv) receives the azygos vein.

Objectives   11; I; 1; (1.3).   11; I; 2; (2.11), (2.12).

0.43	71	E
0.48	90	
0.52	80	

610   The pulmonary trunk

    (i)   is continuous with the conus arteriosus.
    (ii) contributes to the left border of the cardiovascular
        shadow in radiographs.
    (iii) is covered by both fibrous and serous pericardium.
    (iv) is below the aortic knuckle in radiographs.

Objectives   11; I; 1; (1.1), (1.3), (1.5), (1.13).
               11; I; 2; (2.11).

0.52	73	E

611   The left pulmonary artery

    (i)   is on a plane anterior to the left main bronchus
        throughout its course.
    (ii) is connected to the aortic arch by the ligamentum
        arteriosum.

(iii) is shorter than the right pulmonary artery.
(iv) passes anterior to the descending aorta.

Objectives   11; I; 1; (1.2), (1.3).
              11; III; 1; (1.1), (1.5).   *0.36*   85   E

612   Features which can be seen inside the right atrium
show that the

(i) wall of the auricle is smooth.
(ii) wall is ridged anteriorly and smooth posteriorly.
(iii) opening of the superior vena cava has a valve.
(iv) opening of the coronary sinus has a valve.

Objectives   11; I; 1; (1.4).   11; I; 2; (2.4).   *0.29*   73   C
                                               0.41   92

613   The left ventricle

(i) is longer and more conical than the right
ventricle.
(ii) is separated from the left lung by the pericardium,
the left phrenic nerve and the pleura.
(iii) forms part of the diaphragmatic surface of the
heart.
(iv) has a septomarginal trabeculum (moderator band).

Objectives   11; I; 1.   —   —   A

614   The left ventricle

(i) has a septomarginal trabeculum (moderator band).
(ii) is more muscular than the right ventricle.
(iii) conveys more blood than the right ventricle.
(iv) does more work than the right ventricle.

Objectives   11; I; 1; (1.5), (1.6).   *0.21*   84   C
                                               0.40   93

615   The interventricular septum

(i)   has muscular and membranous parts.
(ii) bulges into the right ventricle.
(iii) presents trabeculae carneae.
(iv) in part intervenes between the left ventricle
      and the right atrium.

Objectives   11; 1; 1; (1.8).   0.45   56   E
                                                  0.34   84

616   The cardiac apex beat

(i)   is usually located in the 5th left intercostal space.
(ii) is frequently visible.
(iii) lies medial to the mid-clavicular line.
(iv) is the best place for hearing sounds transmitted
      from the aortic valves.

Objectives   11; 1; 1; (1.9), (1.11).   −   −   A

617   In an antero-posterior radiograph of the chest, the
      right border of the mediastinal shadow usually
      includes the

(i)   pulmonary trunk.
(ii) superior vena cava.
(iii) right ventricle.
(iv) right atrium.

Objectives   11; 1; 1; (1.13).   −   −   C

618   In an antero-posterior radiograph of the chest, the
      left border of the mediastinal shadow includes the

(i)   aortic knuckle.
(ii) pulmonary trunk.
(iii) left ventricle.
(iv) left auricle.

Objectives   11; 1; 1; (1,13).   0.52   66   E

K(ii) Blood and nerve supply of the heart, pericardium, lymphatics (Questions 619–627)

619   The stem of the left coronary artery

    (i)   lies posterior to the pulmonary trunk.
    (ii) is anterior to the left atrium.
    (iii) lies beneath the serous pericardium.
    (iv) arises from the left aortic sinus.

Objectives   11; 1; 2; (2.1).   *0.50*   55   E

620   The left coronary artery

    (i)   lies between the pulmonary trunk and left auricle.
    (ii) arises from the ventral aortic sinus.
    (iii) gives off an anterior interventricular branch.
    (iv) has no effective anastomoses with the right
        coronary artery.

Objectives   11; 1; 2; (2.1), (2.5).   –   –   B

621   The anterior interventricular branch of the left coronary artery

    (i)   supplies most of the interventricular septum.
    (ii) descends towards the apex of the heart.
    (iii) turns onto the diaphragmatic surface of the heart.
    (iv) generally anastomoses with the posterior
        interventricular branch of the right coronary
        artery.

Objectives   11; 1; 2; (2.2), (2.3).   –   –   E

622   The right coronary artery

    (i)   arises from the left aortic sinus.
    (ii) supplies the sinu-atrial node.
    (iii) runs posterior to the pulmonary trunk.
    (iv) supplies the right atrium.

Objectives   11; 1; 1; (1.3).   *0.38*   91   C
                11; 1; 2; (2.1), (2.3), (2.8).   *0.55*   95

623 Blood from the myocardium returns to the right
atrium *directly* via the

(i) coronary sinus.
(ii) venae cordis minimae.
(iii) anterior cardiac veins.
(iv) oblique vein of the left atrium.

Objectives 11; 1; 2; (2.4).   —   —   A

624 The atrioventricular bundle

(i) bifurcates at the upper border of the muscular
part of the interventricular septum.
(ii) usually traverses the posterior margin of the
membranous part of the interventricular septum.
(iii) arises from the atrioventricular node.
(iv) is modified cardiac muscle.

Objectives 11; 1; 2; (2.8).   0.40   48   E
                                      0.40   64

625 The atrioventricular node

(i) consists of specialized muscle cells.
(ii) is continuous with the sinu-atrial node via
conducting tissue.
(iii) is usually supplied by the right coronary artery.
(iv) is the 'pacemaker' for cardiac contraction.

Objectives 11; 1; 2; (2.8), (2.9).   0.44   34   B

626 The atrioventricular node lies

(i) within the interatrial septum.
(ii) posterior to the opening of the coronary sinus.
(iii) above the opening of the coronary sinus.
(iv) within the interventricular septum.

Objectives 11; 1; 2; (2.8).   0.27   68   B
                                      0.33   75

627 The fibrous pericardium

(i)   is the outer layer of pericardium.
(ii) blends with the tunica adventitia of the great vessels.
(iii) is fused inferiorly to the central tendon of the diaphragm.
(iv) encloses the whole heart.

Objectives   11; I; 2; (2.11), (2.12).   *0.31*   77   E
                                                       0.40   73

K(iii) Respiratory system (Questions 628–640)

628 In the living body, the trachea

(i)   lies in the mid-line in the thoracic inlet.
(ii) moves down with inspiration.
(iii) divides in the posterior mediastinum.
(iv) deviates to the left near its lower end.

Objectives   11; III; 1; (1.1).   11; III; 2; (2.6), (2.9).   *0.53*   67   A

629 The left main bronchus

(i)   is about 5 cm in length.
(ii) is more in line with the trachea than the right main bronchus.
(iii) is slightly narrower than the right main bronchus.
(iv) gives direct origin to the lingular bronchus.

Objectives   11; III; 1; (1.1), (1.4).   –   –   B

630 The right main bronchus is

(i)   shorter than the left.
(ii) narrower than the left.
(iii) more in line with the trachea than the left main bronchus.
(iv) enclosed by complete cartilaginous rings.

Objectives   11; III; 1; (1.1).   –   –   B

631 In the root of the right lung

(i)   the upper lobe bronchus is superiorly placed.
(ii) the superior pulmonary vein is the most superior structure.
(iii) the pulmonary veins are anteriorly placed.
(iv) the main bronchus is the most anterior structure.

Objectives   11; III; 1; (1.1), (1.3), (1.5).   0.21   91   B
                                                        0.49   87

632 The root of each lung

(i)   perforates the mediastinal pleura.
(ii) contains pulmonary veins anteriorly.
(iii) contains lymph vessels but no nodes.
(iv) contains the main bronchus posteriorly.

Objectives   11; III; 1; (1.1), (1.3), (1.5), (1.10).   0.38   59   C
              11; III; 2; (2.1), (2.2).   0.20   64

633 The hilus of the right lung contains

(i)   the right upper lobe bronchus most superiorly.
(ii) the right bronchial artery.
(iii) the right pulmonary veins antero-inferiorly.
(iv) broncho-pulmonary lymph nodes.

Objectives   11; III; 1; (1.1), (1.3), (1.5), (1.10).   0.42   72   E
                                                         0.55   84

634 The cervical pleura

(i)   is a subdivision of the visceral pleura.
(ii) forms a dome-like projection into the neck through the thoracic inlet.
(iii) passes in front of the subclavian artery.
(iv) can be outlined on the surface by a curved line which ascends 2.5 cm above the medial third of the clavicle.

Objectives   11; III; 2.   0.22   66   C
                                                         0.30   76
                                                         0.24   48

635   Surface projections of lines of pleural reflection cross

  (i)   the left 3rd costal cartilage at its junction with
        the sternum.
  (ii) below the 12th rib.
  (iii) behind the lateral third of the clavicle.
  (iv) behind the sternoclavicular joint.

  Objectives   11; III; 2; (2.1), (2.5).   —   —   C

636   Parietal pleura

  (i)   is continuous with visceral pleura at the root of
        the lung.
  (ii) extends downwards as a recess between the
        diaphragm and lateral chest wall.
  (iii) extends below the medial part of the 12th rib.
  (iv) extends about 2.5 cm above the neck of the 1st
        rib.

  Objectives   11; III; 2; (2.1), (2.2), (2.5).   0.62   58   A

637   In relation to the pleura,

  (i)   continuities between visceral and parietal pleura
        are established at the costo-mediastinal reflection.
  (ii) the costo-diaphragmatic reflection crosses the
        10th rib in the mid-clavicular line.
  (iii) visceral pleura bridges the pulmonary fissures.
  (iv) central diaphragmatic pleura is supplied by the
        phrenic nerves.

  Objectives   11; III; 2; (2.1–2.5).   0.43   45   D
                                                    0.51   52
                                                    0.43   75

638   The surface projection of the pleura extends beyond
      the bony thoracic wall

  (i)   above the anterior part of the first rib.
  (ii) below the costal margin in the mid-axillary line.

(iii) below the neck of the 12th rib.
(iv) above the neck of the 1st rib.

Objectives   11; III; 2; (2.5).   *0.35*   48   B
                                                                    0.31   48
                                                                    0.53   81

639 Pain from the pleura may be referred to the

(i)   chest wall.
(ii) shoulder region.
(iii) anterior abdominal wall.
(iv) neck.

Objectives   11; III; 2; (2.4).   –   –   E

640 A 'valvular' communication between a small bronchus
     and the right pleural cavity may produce

(i)   mediastinal shift to the left.
(ii) impaired filling of the right atrium.
(iii) depression of the diaphragm on the right side.
(iv) hyper-resonance on the right side.

Objectives   11; III; 2; (2.7), (2.10).   *0.42*   72   E

Chapter 7

The Gastro-intestinal and Genito-urinary Systems

Type A items (Hubbard and Clemans, 1961) (Questions 641–715)

These involve choosing one correct answer from five available choices. The instruction for these items is as follows:

Each of the incomplete statements below is followed by five suggested answers or completions. Select the one which is best in each case.

A(i) Gastro-intestinal tract and peritoneum (Questions 641–659)

641   The oesophagus

    (A) commences at the level of the 4th thoracic vertebra.
    (B) pierces the central tendon of the diaphragm.
    (C) has a skeletal muscle layer throughout its length.
    (D) is crossed on its left anterior aspect by the arch of the aorta.
    (E) is about 12 cm in length.

    Objectives   12; I; 1; (1.1), (1.2), (1.3).   —   —   D

642   The cervical oesophagus

    (A) commences at the level of the upper border of the thyroid cartilage.

(B) lies anterior to the thyroid gland.
(C) lies anterior to the trachea.
(D) has striated muscle in its wall.
(E) has none of the above properties.

Objectives   12; I; 1; (1.1), (1.2), (1.3).   *0.60*   89   D
                                                     0.65   57

643   The oesophagus

(A) lies posterior to the thoracic duct.
(B) has a functional sphincter mechanism at its
    lower end.
(C) enters the stomach in the xiphisternal plane.
(D) is separated from the left atrium by the
    descending aorta.
(E) is not pain-sensitive.

Objectives   12; I; 1; (1.1—1.5).   —   —   B

644   The oesophagus

(A) begins at the upper border of the thyroid
    cartilage.
(B) enters the thorax in the mid-line.
(C) measures about 40 cm.
(D) crosses anterior to the thoracic aorta.
(E) has none of the above properties.

Objectives   12; I; 1; (1.1), (1.2).   —   —   D

645   The oesophagus is in direct relation to the vertebral
      bodies from the

(A) cricoid cartilage to the median arcuate ligament.
(B) cricoid cartilage to the oesophageal hiatus of the
    diaphragm.
(C) cricoid cartilage to the lower limit of the superior
    mediastinum.
(D) thoracic inlet to the oesophageal hiatus of the
    diaphragm.

(E) thoracic inlet to the lower limit of the superior
mediastinum.

Objectives   12; 1; 1; (1.1), (1.2).   *0.37*   47   (

646   The oesophagus is closely related to the

(A) right sympathetic trunk.
(B) right main bronchus.
(C) medial arcuate ligament.
(D) left phrenic nerve.
(E) pericardium.

Objectives   12; 1; 1; (1.2).   *0.56*   68   F

647   The oesophagus

(A) passes between the crura of the diaphragm.
(B) pierces the central tendon of the diaphragm.
(C) pierces the left crus of the diaphragm.
(D) is a posterior relation of the left atrium.
(E) has none of the above properties.

Objectives   12; 1; 1; (1.1), (1.2).   *0.52*   37   I

648   The oesophagus

(A) commences about 25 cm from the incisor teeth
in the average adult.
(B) is constricted to some extent by the right main
bronchus.
(C) is anterior to the thoracic aorta above the
diaphragm.
(D) usually passes between the two crura of the
diaphragm.
(E) has an anatomical sphincter at its lower end just
below the diaphragm.

Objectives   12; 1; 1; (1.1), (1.2), (1.4).   —   —   (

649   The oesophagus

(A) often has a constriction where it is crossed by
the right main bronchus.
(B) contains skeletal muscle throughout the length
of its wall.
(C) is drained by systemic veins only.
(D) crosses posterior to the aorta to reach its left
side.
(E) receives its motor innervation via the vagus nerves.

Objectives   12; I, 1; (1.1–1.5), (5.5).   *0.46*   79   E

650   A surface landmark which constitutes a guide to
the gastro-oesophageal orifice is the

(A) 7th left costal cartilage.
(B) left linea semilunaris.
(C) tip of the 9th left costal cartilage.
(D) 8th thoracic vertebra.
(E) left nipple.

Objectives   12; I; 8; (8.1).   —   —   A

651   The duodenum

(A) is entirely retroperitoneal.
(B) is mainly retroperitoneal.
(C) is related posteriorly to the left ureter.
(D) crosses anterior to the superior mesenteric
artery.
(E) has none of the above properties.

Objectives   12; I; 4; (4.1), (4.2).   *0.64*   82   B
                                                    0.43   92
                                                    0.41   80

652   The horizontal part of the duodenum lies anterior to

(A) the coeliac trunk.
(B) the portal vein.

(C) the inferior vena cava.
(D) the left suprarenal gland.
(E) none of the above.

Objectives   12; 1, 4.

0.30	71	C
0.45	79	
0.32	76	

653 The left colic flexure

(A) lies lower than the right colic flexure.
(B) is supplied by the middle colic artery.
(C) receives parasympathetic vagal fibres.
(D) lies below the left kidney.
(E) has none of the above properties.

Objectives   12; 1; 4.

–	–	E

654 The caecum

(A) has a complete coat of longitudinal muscle.
(B) is retroperitoneal.
(C) has taeniae coli.
(D) is supplied by the middle colic artery.
(E) has none of the above properties.

Objectives   12; 1; 4; (4.5), (5.1).

0.58	84	C
0.22	88	
0.38	73	

655 The lesser omentum

(A) contains the splenic artery.
(B) contains the right gastro-epiploic artery.
(C) contains the left gastric artery.
(D) is attached to the fissure for the ligamentum
    teres.
(E) has none of the above properties.

Objectives   12; 1; 2; (2.2); 5; (5.1).

0.28	26	C

656   The gastrolienal ligament contains

   (A) the splenic artery.
   (B) the tail of the pancreas.
   (C) the short gastric arteries.
   (D) the left gastric artery.
   (E) none of the above.

   Objectives   12; 1; 2; (2.1), (2.2); 5; (5.1).   *0.51*   70   C
                                                          0.58   60

657   The rectum

   (A) is devoid of peritoneum.
   (B) is surrounded by peritoneum.
   (C) has peritoneum on the front throughout its
       length.
   (D) has peritoneum on the front and sides of its
       upper part.
   (E) has peritoneum on its anterior surface only.

   Objectives   12; 1; 4; (4.5).   *0.24*   58   D
                                                          0.39   92

658   The rectum is covered anteriorly by peritoneum

   (A) throughout its length.
   (B) for its upper two-thirds.
   (C) for its upper half.
   (D) for its upper third.
   (E) for none of the above proportions of its length.

   Objectives   12; 1; 4; (4.5).   —   —   B

659   Structures which are retroperitoneal include the

   (A) sigmoid colon.
   (B) caecum.
   (C) appendix.
   (D) head of the pancreas.
   (E) tail of the pancreas.

   Objectives   12; 1; 4; (4.5); 7; (7.4).   *0.37*   95   D

A(ii) Liver, pancreas and spleen (Questions 660–668)

660   The bile duct

   (A) is formed by the junction of the right and left
        hepatic ducts.
   (B) lies behind the portal vein.
   (C) lies in the free edge of the lesser omentum.
   (D) is 2.5 cm in length.
   (E) has none of the above properties.

   Objectives   12; 1; 6; (6.6), (6.9).   *0.45*   92   C

661   The bile duct

   (A) lies in the greater omentum.
   (B) opens into the descending part of the duodenum
        with the pancreatic duct.
   (C) lies in front of the descending part of the
        duodenum.
   (D) is formed by the union of the right and left
        hepatic ducts.
   (E) has none of the above properties.

   Objectives   12; 1; 6, (6.9).   *0.41*   93   B

662   The bile duct

   (A) is formed by junction of the hepatic ducts.
   (B) is posterior to the portal vein.
   (C) opens into the descending part of the duodenum
        with the pancreatic duct.
   (D) opens into the duodenum proximal to the
        pancreatic duct.
   (E) crosses anterior to the upper part of the duodenum.

   Objectives   12; 1; 6; (6.9).   *0.53*   96   C

663   The bile duct

   (A) is in direct contact with the inferior vena cava.

(B) is formed by the union of right and left hepatic
ducts.
(C) runs anterior to the upper part of the duodenum.
(D) runs in the free edge of the lesser omentum
anterior to the portal vein.
(E) commences at the neck of the gall bladder.

Objectives   12; 1; 6; (6.9).   *0.44*   92   D

664   The pancreas

(A) secretes via a duct or ducts into the superior part
of the duodenum.
(B) has a neck behind which the portal vein is
formed.
(C) possesses a tail which is normally contained in
the gastrolienal ligament.
(D) is not related to lesser sac peritoneum.
(E) receives its arterial blood from branches of the
coeliac trunk and inferior mesenteric artery.

Objectives   12; 1; 5; (5.2); 7; (7.1), (7.2), (7.4).   *0.66*   71   B

665   The pancreas usually receives branches from the

(A) ascending left colic artery.
(B) splenic artery.
(C) right colic artery.
(D) left gastric artery.
(E) none of the above.

Objectives   12; 1; 5; (5.2).   —   —   B

666   The main pancreatic duct

(A) opens into the duodenum proximal to the
accessory duct.
(B) joins the bile duct at the apex of the greater
duodenal papilla.
(C) usually communicates with the accessory
pancreatic duct when this is present.

(D) drains the endocrine secretion of the body of
the pancreas.
(E) opens into the superior part of the duodenum.

Objectives   12; I; 4; (4.1); 7; (7.1), (7.2).   *0.38*   59   C

667   The tail of the pancreas

(A) lies in the gastrolienal ligament.
(B) lies in the lienorenal ligament.
(C) lies posterior to the spleen.
(D) is retroperitoneal.
(E) has none of the above properties.

Objectives   12; I; 7; (7.1), (7.4).   —   —   B

668   The spleen

(A) extends forwards to the left costal margin.
(B) receives its main blood supply via the
gastrolienal ligament.
(C) sends venous drainage into the left renal vein.
(D) lies within the lesser sac.
(E) is separated by peritoneum from the diaphragm.

Objectives   12; I; 3; (3.8), (3.10); 5; (5.1), (5.4).   *0.48*   59   E

A(iii) Genito-urinary system (Questions 669–693)

669   The left kidney (with its fascial coverings) is

(A) in direct contact with the pancreas.
(B) lower than the right kidney.
(C) not in contact with peritoneum.
(D) in direct contact with the duodenum.
(E) in contact posteriorly with the 11th and 12th ribs.

Objectives   12; II; 1; (1.4), (1.6).   *0.38*   69   A

570   The ureters are

(A) 6 cm from the cervix uteri.
(B) 2 cm from the cervix uteri.
(C) on a plane anterior to the renal vessels.
(D) on a plane anterior to the right and left colic
     vessels.
(E) crossed inferiorly by the uterine arteries.

Objectives   12; II; 1.   *0.20*   32   B

671   The terminal part of the ductus deferens

(A) lies lateral to the seminal vesicle.
(B) is ampullated.
(C) is covered by peritoneum.
(D) lies behind the prostate.
(E) has none of the above properties.

Objectives   12; III; 1.   —   —   B

672   The ductus deferens is connected to the prostatic
      urethra by

(A) the prostatic utricle.
(B) the prostatic sinus.
(C) the ejaculatory duct.
(D) the urachus.
(E) none of the above.

Objectives   12; III; 1; (1.6), (1.10).   —   —   C

673   The seminal vesicles

(A) separate the terminal parts of the ureters from
     contact with the bladder wall.
(B) are partially invested by the peritoneum of the
     rectovesical pouch.
(C) freely communicate with each other across the
     mid-line.

    (D) terminate in ejaculatory ducts which enter the
        apex of the prostate.

    (E) have none of the above properties.

Objectives   12; II; 2; (2.3).   12; III; 1; (1.10).   *0.35*   52   B

674   The seminal vesicles

    (A) lie medial to the termination of each ductus deferens.

    (B) are anterior relations of the prostate gland.

    (C) are posterior relations of the bladder.

    (D) empty directly into the prostatic urethra.

    (E) have none of the above properties.

Objectives   12; II; 2; (2.2).   12; III; 1; (1.10).   –   –   C

675   Each seminal vesicle

    (A) opens by numerous ducts into the prostatic sinus.

    (B) opens into the spongy urethra.

    (C) can be felt through the posterior wall of the
        rectum.

    (D) lies lateral to the ampulla of the ductus deferens.

    (E) lies on the superior surface of the bladder.

Objectives   12; II; 2; (2.3), (2.6).   *0.39*   77   D
              12; III; 1; (1.10).   *0.41*   81

676   The ejaculatory ducts

    (A) are formed by the union of the prostatic ducts
        and ducts of the seminal vesicle.

    (B) lie on the superior surface of the bladder.

    (C) open into the membranous urethra.

    (D) open into the bladder.

    (E) have none of the above properties.

Objectives   12; II; 2; (2.2), (2.3), (2.6).   –   –   E
              12; III, 1; (1.8), (1.10).

677   The ejaculatory ducts drain into

   (A) the bladder.
   (B) the spongy portion of the urethra.
   (C) the seminal vesicles.
   (D) the prostatic utricle.
   (E) none of the above.

   Objectives   12; II; 2; (2.6).   12; III; 1; (1.10).   −'   −   E

678   The prostate gland

   (A) lies below the neck of the bladder.
   (B) is situated outside the pelvis.
   (C) lies in the urogenital diaphragm.
   (D) opens via a single duct into the membranous
        urethra.
   (E) has none of the above properties.

   Objectives   12; II; 2; (2.3), (2.6).   *0.33*   86   A
                12; III; 1; (1.8), (1.9).   *0.41*   89

679   The prostate gland

   (A) lies directly anterior to the seminal vesicles.
   (B) has multiple ducts opening into the membranous
        urethra.
   (C) lies in the urogenital diaphragm.
   (D) opens via a single duct into the urethra.
   (E) has none of the above properties.

   Objectives   12; II; 2; (2.3), (2.6).   *0.41*   89   E
                12; III; 1; (1.8), (1.9).   *0.48*   62

680   The prostatic urethra

   (A) is 5 cm in length.
   (B) lies near the posterior surface of the prostate.
   (C) contains a crest on which the orifices of the
        prostatic ducts open.

(D) shows prostatic sinuses on either side of the
     openings of the ejaculatory ducts.
(E) has none of the above properties.

Objectives   12; II; 2; (2.6).   12; III; 1; (1.8), (1.10). *0.47*   72   D

681 The prostate gland

(A) lies in the superficial perineal pouch.
(B) lies in the deep perineal pouch.
(C) lies around the membranous part of the urethra.
(D) has a peritoneal covering posteriorly.
(E) has none of the above properties.

Objectives   12; II; 2; (2.6).   12; III; 1; (1.8), (1.10).   –   –   E

682 The prostate gland

(A) is anteriorly related to peritoneum.
(B) is posteriorly related to peritoneum.
(C) is anteriorly related to the bladder.
(D) is separated from the bladder by the seminal
     vesicles.
(E) has none of the above properties.

Objectives   12; II; 2; (2.3).   12; III; 1; (1.8), (1.10).   –   –   E

683 The narrowest part of the male urethra is

(A) at the external urethral orifice.
(B) at the junction between spongy and membranous
     urethra.
(C) in the most proximal part of the membranous
     urethra.
(D) in the prostatic urethra at the level of the
     colliculus seminalis.
(E) at the internal urethral orifice.

Objectives   12; II; 2; (2.6).   12; III; 1; (1.8).   *0.39*   63   A

684   The epididymis

(A) is enclosed by the tunica vaginalis.
(B) has a sinus along the medial surface of the testis.
(C) receives its main blood supply via the artery of
the ductus deferens.
(D) has its major lymphatic drainage along the
testicular artery to lumbar nodes.
(E) does not contribute to the volume of seminal
fluid.

Objectives   12; III; 1.   *0.35*   55   D

685   The primary role in the support of the uterus is
performed by the

(A) broad ligaments.
(B) levatores ani muscles.
(C) round ligaments of the uterus.
(D) transverse cervical ligaments.
(E) uterosacral ligaments.

Objectives   12; IV; 1; (1.6).   —   —   B

686   The ligament most important in retaining the
position of the uterus is

(A) the broad ligament.
(B) the transverse cervical ligament.
(C) the round ligament of the uterus.
(D) the utero-sacral ligament.
(E) none of the above.

Objectives   12; IV; 1; (1.6).   —   —   B

687   The uterine tube

(A) does not open into the peritoneal cavity.
(B) lies in the upper border of the broad ligament
along its whole extent.
(C) has a laterally situated infundibulum.

(D) lies within the mesovarium.
(E) has fimbriae attached to its ampulla.

Objectives   12; IV; 1; (1.1), (1.2), (1.5).   0.40   64   C
                                                     0.45   64
                                                     0.21   71

688   The ovaries

(A) lie between the right and left common iliac
    arteries.
(B) lie between the internal iliac vessels and the
    ureters.
(C) lie on the psoas major muscles.
(D) are suspended by the round ligaments.
(E) have none of the above properties.

Objectives   12; IV; 1; (1.1).   0.40   37   E

689   The ovary is closely related to the

(A) sacro-iliac joint.
(B) psoas major muscle.
(C) obturator nerve.
(D) trunk of the uterine artery.
(E) round ligament.

Objectives   12; IV; 1.   0.50   49   C

690   The vagina

(A) usually forms an angle of 60 degrees with the
    axis of the uterus.
(B) is related anteriorly to the peritoneum.
(C) is a postero-inferior relation of the ureter.
(D) has a longer anterior than posterior wall.
(E) has none of the above properties.

Objectives   12; IV; 1; (1.1), (1.5), (1.7).   0.25   44   C

691 The vagina

    (A) has mucous glands in its wall which are principally responsible for lubrication during coitus.

    (B) lies in a plane about 60 degrees from the horizontal.

    (C) is the most erotogenic of the female genitalia.

    (D) is directly related to peritoneum over its anterior fornix.

    (E) is entered by the cervix uteri on its superior aspect.

Objectives   12; IV; 1; (1.1), (1.5), (1.7).   *0.33*   69   B

692 The vagina

    (A) is not in contact with peritoneum.

    (B) is related to the peritoneum of the recto-uterine pouch.

    (C) is related to the peritoneum of the utero-vesical pouch.

    (D) lies anterior to the urethra.

    (E) has a longer anterior than posterior wall.

Objectives   12; IV; 1; (1.1), (1.5), (1.7).   *0.35*   56   B
                                                              0.39   85

693 The female urethra

    (A) traverses the clitoris.

    (B) opens anterior to the clitoris.

    (C) opens between vagina and anus.

    (D) is fused with the anterior wall of the vagina.

    (E) has none of the above properties.

Objectives   12; II; 2; (2.5).   12; IV; 1; (1.8).   *0.30*   87   D

A(iv) Blood supply, lymphatic drainage and innervation (Questions 694–715)

694 The arterial supply of the stomach is from branches of the

    (A) left gastric artery.

    (B) left gastric and splenic arteries.

(C) left gastric, splenic and common hepatic arteries.
(D) left gastric, splenic and superior mesenteric arteries.
(E) coeliac and superior mesenteric arteries.

Objectives　12; I; 5; (5.1), (5.2).　　　　0.37　77　C

695　Branches of the superior mesenteric artery include

(A) left colic.
(B) superior pancreatico-duodenal.
(C) ileocolic.
(D) hepatic.
(E) gastroduodenal.

Objectives　12; I; 5; (5.1), (5.2).　　　　0.70　93　C

696　The superior mesenteric artery

(A) commences at the inferior border of the horizontal part of the duodenum.
(B) supplies the entire small intestine.
(C) has branches which anastomose with branches of both the coeliac trunk and the inferior mesenteric artery.
(D) gives rise to the superior pancreatico-duodenal artery.
(E) has none of the above properties.

Objectives　12; I; 5; (5.1), (5.2), (5.3).　　　–　–　C

697　The middle colic artery is a branch of the

(A) superior mesenteric artery.
(B) inferior mesenteric artery.
(C) coeliac trunk.
(D) common hepatic artery.
(E) left gastric artery.

Objectives　12; I; 5; (5.1), (5.2).　　　　0.95　98　A

698 The marginal artery of the large intestine is formed
     in part by branches of the

   (A) splenic artery.
   (B) left gastro-epiploic artery.
   (C) coeliac trunk.
   (D) inferior mesenteric artery.
   (E) common hepatic artery.

   Objectives   12; I; 5; (5.1–5.3).   *0.69*   98   D

699 The rectum

   (A) receives its blood supply from the inferior
       mesenteric artery only.
   (B) is covered by peritoneum anteriorly in its upper
       one-third only.
   (C) is about 5 cm in length.
   (D) commences at the upper end of the sacrum.
   (E) has the bladder directly anterior to it in the male.

   Objectives   12; I; 4; (4.5), (4.8); 5; (5.1), (5.2).   *0.44*   57   E

700 The renal artery

   (A) lies anterior to the renal vein.
   (B) has no branches except to the kidney.
   (C) is higher on the right than on the left side of
       the body.
   (D) lies anterior to the pelvis of the ureter.
   (E) is the only paired branch of the aorta.

   Objectives   12; II; 1.   *0.49*   80   D

701 The portal vein

   (A) receives blood from the kidneys.
   (B) lies in the greater omentum.
   (C) is formed by junction of the superior and inferior
       mesenteric veins.

(D) commences behind the pancreas.
(E) has none of the above properties.

Objectives   12; 1, 5; (5.4).   *0.34*   90   D

702   The portal vein

(A) usually lies anterior and to the right of the bile
duct.
(B) may communicate directly with the inferior vena
cava.
(C) has numerous valves.
(D) receives blood only from the alimentary tract.
(E) usually commences behind the neck of the
pancreas.

Objectives   12; 1; 5; (5.4).   *0.48*   82   E

703   The portal vein is formed by the union of

(A) inferior mesenteric and superior mesenteric veins.
(B) hepatic and splenic veins.
(C) left and right gastric veins.
(D) splenic and superior mesenteric veins.
(E) none of the above combinations.

Objectives   12; 1; 5; (5.4).   *0.47*   91   D
                                                  0.49   92
                                                  0.40   92

704   The splenic vein

(A) drains into the inferior vena cava.
(B) drains into the left renal vein.
(C) unites with the inferior mesenteric vein to form
the portal vein.
(D) receives the left testicular vein.
(E) joins the superior mesenteric vein.

Objectives   12; 1; 5; (5.4).   *0.46*   78   E

705   The anal canal

  (A) is supplied by the superior rectal vessels.
  (B) sends lymphatics to the external iliac lymph
       nodes.
  (C) presents the pecten above the anal valves.
  (D) is covered by peritoneum on its lateral surfaces.
  (E) has none of the above properties.

Objectives   12; I; 4; (4.5); 5; (5.1), (5.2).
              12; V; (1.4), (1.6).   —   —   A

706   The right renal vein receives

  (A) the gonadal vein.
  (B) the inferior phrenic vein.
  (C) the suprarenal vein.
  (D) the inferior mesenteric vein.
  (E) none of the above.

Objectives   12; II; 1.   —   —   E

707   The right testicular vein usually drains into the

  (A) inferior vena cava.
  (B) right renal vein.
  (C) portal vein.
  (D) inferior vena cava by a common trunk which is
       shared with the left testicular vein.
  (E) none of the above.

Objectives   12; V; 1; (1.6).   —   —   A

708   The small intestine

  (A) has lymphatic vessels which mainly run at right-
       angles to the long axis of its wall.
  (B) has a thinner wall in its jejunal than its ileal part.
  (C) is suspended throughout by the mesentery.
  (D) has circular folds of mucosa throughout its
       extent.
  (E) is supplied by the coeliac plexus only.

Objectives   12; I; 4; (4.3), (4.4); 5; (5.8), (5.9).   *0.46*   61   A

709   The small intestine

(A) has no lymph nodes in its mesentery.
(B) contains intramural lymphatic tissue in the
     ileum only.
(C) drains lymph directly into the thoracic duct.
(D) is drained by lymph vessels that follow the blood
     vessels of the small intestine.
(E) has none of the above properties.

Objectives   12; I; 4; (4.4); 5; (5.8).   —   —   D

710   Lymphatics of the body of the testis most
      commonly drain into

(A) superficial inguinal lymph nodes.
(B) deep inguinal lymph nodes.
(C) lumbar lymph nodes.
(D) external iliac lymph nodes.
(E) none of the above.

Objectives   12; V; 1; (1.6).   *0.45*   69   C
                                                   0.62   88
                                                   0.48   94

711   The greater splanchnic nerve contains mainly

(A) somatic afferent fibres.
(B) somatic efferent fibres.
(C) pre-ganglionic sympathetic fibres.
(D) post-ganglionic sympathetic fibres.
(E) parasympathetic fibres from the vagus.

Objectives   12; I; 5; (5.9).   *0.31*   44   C

712   The coeliac ganglion

(A) is concerned with the parasympathetic
     innervation of the gut.
(B) gives post-ganglionic motor fibres to the
     suprarenal medulla.

(C) contains the ganglion cells of visceral afferent neurons.
(D) gives post-ganglionic fibres to the stomach.
(E) lies in the mid-line below the coeliac trunk.

Objectives   12; I; 5; (5.9).   *0.50*   51   D

713 Parasympathetic innervation of the pelvic viscera involves

(A) the vagus.
(B) the greater splanchnic nerves.
(C) the pelvic splanchnic nerves.
(D) the inferior mesenteric plexus.
(E) somatic nerves.

Objectives   12; I; 5; (5.9).   12; V; 1; (1.5).   *0.64*   86   C

714 The sigmoid colon

(A) receives parasympathetic innervation via the vagus.
(B) sends venous drainage into the inferior vena cava.
(C) is retroperitoneal.
(D) is supplied by the pelvic splanchnic nerves.
(E) is within the vascular territory of the superior mesenteric artery.

Objectives   12; I; 5; (5.1), (5.2), (5.4), (5.9).
             12; V; (1.5).   *0.47*   81   D

715 The splenic flexure

(A) is drained by the splenic vein.
(B) is supplied mainly by the superior mesenteric artery.
(C) is innervated mainly by the vagus.
(D) is situated at a lower level than the hepatic flexure.
(E) has none of the above properties.

Objectives   12; I; 4; (4.5); 5; (5.2), (5.4), (5.9).   —   —   E

Type E items (Hubbard and Clemans, 1961) (Questions 716–737)

These items are of the assertion—reason type and there are five possible responses. The instruction for these items is as follows:

Each question consists of an assertion and a reason. Responses should be chosen as follows:
- *(A) if the assertion and reason are true statements and the reason is a correct explanation of the assertion;*
- *(B) if the assertion and reason are true statements but the reason is NOT a correct explanation of the assertion;*
- *(C) if the assertion is true but the reason is a false statement;*
- *(D) if the assertion is false but the reason is a true statement;*
- *(E) if both assertion and reason are false statements.*

E(i) Gastro-intestinal tract and peritoneum (Questions 716–721)

716   In the cadaver the first part of the duodenum is often stained green

BECAUSE

the gall bladder is in contact with the superior part of the duodenum.

Objectives	12; 1; 4; (4.1), (4.2); 6; (6.9).	0.35	65	A
		0.44	59	
		0.39	71	

717   The jejunum has a greater absorptive area than the ileum

BECAUSE

it has more circular folds and longer villi.

Objectives	12; 1; 4; (4.3), (4.4).	0.19	83	A

718   The small intestine is the most mobile part of the alimentary tract

BECAUSE

it is suspended by the greater omentum.

Objectives	12; 1; 2; (2.2); 4; (4.1), (4.3).	0.46	81	C
		0.37	72	
		0.34	83	

719   On barium meal examination the appearance of the
      jejunum is more feathery than that of the ileum

      BECAUSE

      there are more villi in the jejunum than the ileum.

      Objectives   12; I; 4; (4.3), (4.10).   —   —   B

720   A barium meal radiograph of the jejunum shows
      numerous sacculations

      BECAUSE

      this part of the intestine has an incomplete coat
      of longitudinal muscle.

      Objectives   12; I; 4; (4.3), (4.4), (4.10).   0.56   65   E
                                                           0.39   75
                                                           0.42   67

721   The vermiform appendix may be located by following
      the taeniae coli inferiorly over the ascending colon
      and caecum

      BECAUSE

      thus traced, the taeniae coli converge to form a
      complete longitudinal coat over the appendix.

      Objectives   12; I; 4; (4.5), (4.7).   —   —   A

E(ii) Liver, pancreas and spleen (Questions 722–725)

722   The right lobe of the liver is much larger than the
      left

      BECAUSE

      its main vascular connections include in their
      distribution the caudate and quadrate lobes.

      Objectives   12; I; 5; (5.1), (5.4); 6; (6.3).   0.34   42   C

723   The liver requires a dual blood supply from the
portal vein and the hepatic artery

BECAUSE

the portal blood enters the liver sinusoids whereas
blood from the hepatic artery is distributed solely
to the connective tissues of the liver.

Objectives   12; I; 5; (5.1), (5.4); 6; (6.4), (6.5).   *0.47*   66   C
0.40   57
0.24   59
0.46   81

724   The pancreas comes into direct relation with the
spleen

BECAUSE

its tail is surrounded by peritoneum between the
layers of the lieno-renal ligament.

Objectives   12; I; 3; (3.8), (3.10); 7; (7.1), (7.4).   *0.51*   44   A

725   Total removal of the spleen from an adult has no
serious lasting effects

BECAUSE

all the functions of the spleen can be carried out
by the liver.

Objectives   12; I; 3; (3.9).   *0.31*   79   C

A(iii) Genito-urinary system (Questions 726–730)

726   The seminal vesicles can normally be felt (palpated)
through the wall of the rectum when the bladder is
empty

BECAUSE

they are immediate posterior relations of the
prostate.

Objectives   12; I; 4; (4.8).   12; III; 1; (1.8), (1.10).   *0.18*   26   E

727 The prostate gland cannot be palpated through the
     wall of the rectum

     BECAUSE

     it is separated from the rectum by the seminal
     vesicles.

     Objectives   12; I; 4; (4.8).   12; III; 1; (1.8), (1.10).   –   –   E

728 The prostate cannot be palpated per rectum

     BECAUSE

     the rectovesical pouch of the peritoneum descends
     to the apex of the prostate.

     Objectives   12; I; (4.8).   12; III; (1.8).   –   –   E

729 The levatores ani muscles are important supports
     of the uterus

     BECAUSE

     the cervix uteri rests directly on them.

     Objectives   12; IV; 1; (1.6).   0.22   43   C

730 The broad ligament forms the most significant
     ligamentous support of the uterus

     BECAUSE

     it attaches the whole body of the uterus to the
     lateral wall of the pelvis.

     Objectives   12; IV; 1; (1.1), (1.6).   0.44   60   D
                                                              0.39   83

A(iv) Blood supply, lymphatic drainage and innervation (Questions 731–737)

731   Accidental ligation of the hepatic artery proper close
      to its point of division would cut off all arterial blood
      to the liver

      BECAUSE

      the distribution of the hepatic artery proper is
      restricted to the liver and there is no other source
      of supply.

      Objectives   12; I; 5; (5.1), (5.2), (5.3); 6; (6.2).   *0.46*   75   C
                                                                    0.51   60
                                                                    0.33   73

732   Liver tissue receives blood both from branches of
      the portal vein and branches of the hepatic artery
      proper

      BECAUSE

      portal-systemic anastomoses directly connect
      branches of the coeliac trunk with tributaries of
      the portal vein.

      Objectives   12; I; 5; (5.1), (5.2), (5.5).   *0.24*   67   C

733   Pain arising from hollow viscera is usually poorly
      localized

      BECAUSE

      visceral pain fibres travel mainly in the parasympathetic
      nervous system.

      Objectives   12; I; 5; (5.9).   *0.37*   85   C
                                                                    0.31   70

734   Irritation of the diaphragmatic peritoneum may
      produce pain in the shoulder

      BECAUSE

      the skin over the shoulder is supplied by the same
      cord segments (C3, C4) as are represented in the
      phrenic nerve.

      Objectives   12; I; 2; (2.3).   *0.32*   85   A

735 Removal of the superior hypogastric plexus interferes
with normal micturition

BECAUSE

micturition reflexes are dependent upon intact
parasympathetic pathways.

Objectives   12; II; 2; (2.8), (2.9).   —   —   D

736 Section of the sympathetic fibres to the bladder
results in incontinence of urine

BECAUSE

the major innervation of the detrusor muscle is
provided by the superior hypogastric plexus.

Objectives   12; II; 2; (2.8), (2.9).   0.59   62   E

737 Removal of the lumbar part of both sympathetic
trunks would effectively provide sympathetic
denervation within the abdomen

BECAUSE

all the pre-ganglionic sympathetic nerve fibres to
abdominal viscera are carried in ventral rami above
the L2 level.

Objectives   12; I; 5; (5.9).   0.39   48   D

Type K items (Hubbard and Clemans, 1961) (Questions 738–799)

These items are in the form of incomplete statements and call for recognition
of one or more correct responses which are grouped in various ways. The
instruction for these items is as follows:

*For each of the incomplete statements below, one or more of the
completions given is correct. Responses should be chosen as follows:*
(A) if only completions (i), (ii) and (iii) are correct,
(B) if only completions (i) and (iii) are correct,
(C) if only completions (ii) and (iv) are correct,
(D) if only completion (iv) is correct,
(E) if all completions are correct.

K(i) Gastro-intestinal tract and peritoneum (Questions 738–759)

738   The oesophagus

(i)   begins at the level of the 6th cervical vertebra.
(ii) is crossed anteriorly by the thoracic duct.
(iii) enters the abdomen at the level of the 11th
      thoracic vertebra.
(iv) is lined by pseudostratified columnar epithelium.

Objectives   12; l; 1; (1.1), (1.2), (1.3).   —   —   B

739   Amongst the chief structures forming the stomach
      bed are the

(i)   pancreas.
(ii) left suprarenal gland.
(iii) left kidney.
(iv) left lobe of the liver.

Objectives   12; l; 3; (3.5).   —   —   A

740   Amongst the chief structures forming the stomach
      bed are the

(i)   transverse mesocolon.
(ii) pancreas.
(iii) spleen.
(iv) right kidney.

Objectives   12; l; 3; (3.5).   0.49   81   A

741   After a barium meal, the stomach

(i)   is lined throughout with barium when the
      subject is in the upright position.
(ii) may show the markings of rugae.
(iii) lies fully above the transtubercular plane when
      the subject is upright.
(iv) shows a pyloric antrum continuously occupied
      by barium.

Objectives   12; l; 3; (3.2), (3.6).   0.26   84   C

742   The duodenum

 (i)   is related to the liver in both its upper and
        descending parts.
 (ii) is retroperitoneal except for its upper part.
 (iii) has a horizontal part crossed by the root of the
        mesentery.
 (iv) has an ascending part on the left of the aorta.

Objectives   12; 1; 4; (4.1), (4.2).   —   —   E

743   The duodenum

 (i)   is in direct contact with the aorta.
 (ii) has mobile upper and ascending parts.
 (iii) often has a lesser duodenal papilla superior and
        anterior to the greater duodenal papilla.
 (iv) receives arterial supply mainly from branches of
        the left gastric and splenic arteries.

Objectives   12; 1; 4; (4.1), (4.2); 5; (5.1), (5.2).   —   —   B

744   The descending part of the duodenum

 (i)   has the opening of the accessory pancreatic
        duct, if present, proximal to the opening of the
        main pancreatic duct.
 (ii) is anterior to the right kidney.
 (iii) is completely retroperitoneal.
 (iv) is characterized by circular mucosal folds.

Objectives   12; 1; 4; (4.1), (4.2).   *0.48*   59   E

745   The jejunum

 (i)   contains more circular folds than the ileum.
 (ii) has more fat in its mesentery than the ileum.
 (iii) is supplied entirely by the superior mesenteric
        artery.
 (iv) contains large aggregated lymphatic follicles.

Objectives   12; 1; 4; (4.3), (4.4); 5; (5.1), (5.2).   *0.36*   75   B
                                                        0.29   85

746 Sphincters are anatomically demonstrable at the

(i) pylorus.
(ii) cardiac orifice.
(iii) anorectal junction.
(iv) duodenojejunal junction.

Objectives   12; I; 1; (1.3); 4; (4.1), (4.3), (4.5), (4.9).   *0.25*   57   B
0.18   52

747 In examining barium meal radiographs in a normal
subject you expect to find

(i) a regular edge to the barium shadow in the
situation of the lesser curvature.
(ii) an irregular barium outline in the situation
of the greater curvature.
(iii) a regular triangular or globular shadow in the
position of the first part of the duodenum.
(iv) a highly irregular barium shadow within the
jejunum.

Objectives   12; I; 3; (3.6); 4; (4.10).   *0.36*   44   E

748 The features which are normally never found in the
ileum include

(i) haustra.
(ii) gas shadows in radiographs.
(iii) appendices epiploicae.
(iv) taeniae coli.

Objectives   12; I; 4; (4.3), (4.10).   —   —   E

749 The colon (excluding caecum) is characterized
throughout by the presence of

(i) appendices epiploicae.
(ii) villi.
(iii) taeniae.
(iv) mesenteries.

Objectives   12; I; 4; (4.5), (4.6).   *0.25*   82   B
0.54   88

750 The caecum

    (i)   has the appendix attached to its antero-medial
        aspect.
    (ii) usually lies in the true pelvis.
    (iii) has anterior and posterior folds guarding the
        ileo-caecal opening.
    (iv) is mainly surrounded by peritoneum.

Objectives   12; I; 4; (4.5).   *0.48*   51   D

751 Structures that can normally be palpated by a digit
    inserted into the rectum of an adult male include the

    (i)   appendix.
    (ii) bladder (when empty).
    (iii) ureters.
    (iv) prostate.

Objectives   12; I; 4; (4.5), (4.8).   12; II; 2; (2.3).
              12; III; 1; (1.8).   —   —   D

752 The peritoneum has a

    (i)   pain-sensitive parietal layer.
    (ii) parietal layer supplied by somatic nerves.
    (iii) visceral layer which is largely insensitive to pain.
    (iv) visceral layer supplied by autonomic nerve
        plexuses.

Objectives   12; I; 2; (2.3).   *0.37*   34   E

753 The lesser sac

    (i)   communicates with the greater sac.
    (ii) lies behind the stomach.
    (iii) partly extends behind the liver.
    (iv) surrounds the spleen.

Objectives   12; I; 2; (2.1), (2.2).   *0.37*   93   A

754 The lesser omentum

(i)   is attached to both stomach and duodenum.
(ii) forms the anterior boundary of the epiploic
       foramen.
(iii) contains the bile duct, the hepatic artery proper
       and the portal vein.
(iv) is continuous with the peritoneum over the liver
       at the porta hepatis.

Objectives   12; I; 2; (2.1), (2.2); 5; (5.1), (5.2),
             (5.4); 6; (6.6).   −   −   E

755 The lesser sac

(i)   has the quadrate lobe of the liver projecting
       into it superiorly.
(ii) lies anterior to the pancreas.
(iii) has a posterior wall formed partly by the
       anterior layer of the lesser omentum.
(iv) has the inferior vena cava posterior to its
       opening.

Objectives   12; I; 2; (2.1), (2.2).   *0.47*   86   C

756 The greater omentum

(i)   is partly continuous with the lieno-renal ligament.
(ii) serves as a fat storage depot.
(iii) is adherent to the transverse colon.
(iv) is attached to the stomach.

Objectives   12; I; 2; (2.1), (2.2).   *0.37*   67   E

757 The falciform ligament

(i)   connects the liver to the diaphragm and anterior
       abdominal wall.
(ii) encloses the ligamentum teres.
(iii) is continuous with the coronary ligament.
(iv) is attached to the umbilicus.

Objectives   12; I; 6; (6.6), (6.7).   *0.40*   45   E

758 In barium meal radiographs the oesophagus may
     appear indented by the

     (i)   arch of the aorta.
     (ii) left bronchus.
     (iii) left atrium.
     (iv) left ventricle.

     Objectives   12; I; 1; (1.6).   —   —   A

759 The oesophagus is

     (i)   a posterior relation of the oblique sinus of the
           pericardium and the left atrium.
     (ii) an anterior relation of the trachea.
     (iii) an anterior relation of the vena azygos.
     (iv) a posterior relation of the thoracic duct.

     Objectives   12; I; 1; (1.1), (1.2).   *0.32* 55   B

K(ii) Liver, pancreas and spleen (Questions 760–769)

760 The right lobe of the liver

     (i)   is separated from the left lobe by the falciform
           ligament.
     (ii) receives all its blood from the right branch of
           the portal vein.
     (iii) has the right suprarenal gland posterior to it.
     (iv) is situated to the right of the gall bladder.

     Objectives   12; I; 6; (6.3), (6.8), (6.9).   *0.22* 40   B

761 A patient is suspected of having cirrhosis of the liver
     resulting in an increased portal venous pressure.
     Following this line of thought it would be reasonable
     to

     (i)   enquire for a history of vomiting of blood.
     (ii) look for enlarged veins in the anterior abdominal
           wall.

(iii) enquire for a history of bleeding from the rectum.
(iv) palpate for evidence of splenic enlargement.

Objectives   12; 1; 5.   *0.43*   32   E

762   The gall bladder

(i)   is directly related to the caudate lobe of the liver.
(ii) underlies the 9th right costal cartilage.
(iii) lies between the layers of the coronary ligament.
(iv) is related to the transverse colon.

Objectives   12; 1; 6; (6.9).   *0.29*   69   C
                                                          0.56   89
                                                          0.20   63

763   The gall bladder has

(i)   the quadrate lobe directly to its left side.
(ii) a fundus covered by peritoneum.
(iii) a neck continuous with the cystic duct.
(iv) glands that secrete bile and mucus.

Objectives   12; 1; 6; (6.9).   −   −   A

764   The pancreas

(i)   is completely invested by peritoneum.
(ii) forms part of the stomach bed.
(iii) opens via a single duct into the descending part of
      the duodenum.
(iv) lies in contact with the anterior surface of the
      left kidney.

Objectives   12; 1; 7; (7.1), (7.2), (7.4).   *0.54*   92   C
                                                          0.49   88
                                                          0.41   92

765 The pancreas

(i)   lies posterior to the lesser sac.
(ii) has an inferior surface covered by peritoneum
      continuous with the lower layer of the
      transverse mesocolon.
(iii) has an anterior surface covered by peritoneum
      of the lesser sac.
(iv) may be partly within the lieno-renal ligament.

Objectives   12; 1; 7; (7.1), (7.2), (7.4).   0.57   67   E

766 The pancreas

(i)   has a neck which lies anterior to the origin of
      the portal vein.
(ii) is related to the lesser but not the greater sac of
      peritoneum.
(iii) has an uncinate process lying behind the superior
      mesenteric vessels.
(iv) has a tail which lies in the gastro-lienal ligament.

Objectives   12; 1; 7; (7.1), (7.2), (7.4).   0.32   38   B

767 The body of the pancreas

(i)   crosses the vertebral column.
(ii) lies immediately below the coeliac trunk.
(iii) lies immediately above the duodeno-jejunal
      flexure.
(iv) is related to both the greater and lesser sacs.

Objectives   12; 1; 7; (7.1), (7.2), (7.4).   —   —   E

768 The spleen is closely related to the

(i)   left flexure of the colon.
(ii) stomach.
(iii) tail of the pancreas.
(iv) left kidney.

Objectives   12; 1; 3; (3.7), (3.10).   0.45   91   E
                                                      0.48   92

769   The spleen is

  (i)   related to the diaphragm under the 9th to 11th
        left ribs.
  (ii) related to the stomach across the greater sac.
  (iii) directly related to the tail of the pancreas.
  (iv) related to the left colic flexure across the
        greater sac.

  Objectives   12; I; 3; (3.7), (3.10).   —   —   E

K(iii) Genito-urinary system (Questions 770–786)

770   The immediate relations of the right kidney include
      the

  (i)   right suprarenal gland.
  (ii) bile duct.
  (iii) right colic flexure of the colon.
  (iv) neck of the pancreas.

  Objectives   12; II; 1; (1.1), (1.6).   0.23   69   B
                                                            0.19   81

771   Direct anterior relations of the right kidney include the

  (i)   right colic flexure.
  (ii) head of pancreas.
  (iii) right suprarenal gland.
  (iv) bile duct.

  Objectives   12; II; 1; (1.1), (1.6).   0.33   70   B
                                                            0.36   75

772   Direct anterior relations of the right kidney include the

  (i)   right colic flexure.
  (ii) inferior vena cava.
  (iii) right suprarenal gland.
  (iv) ureter.

  Objectives   12; II; 1; (1.1), (1.6).   0.41   88   B

773   The right kidney is related to the

    (i)   right psoas major muscle.
    (ii)   right suprarenal gland.
    (iii)   visceral surface of the liver.
    (iv)   small intestine.

    Objectives   12; II; 1; (1.1), (1.6).   *0.40*   65   E

774   Structures lying against the anterior surface of the
left kidney include the

    (i)   liver.
    (ii)   pancreas.
    (iii)   duodenum.
    (iv)   spleen.

    Objectives   12; II; 1; (1.1), (1.6).   *0.29*   81   C
                                                  0.22   84

775   Structures relevant to the interpretation of the
ureteric shadows in an intravenous pyelogram
include the

    (i)   tips of lumbar transverse processes.
    (ii)   sacro-iliac joint shadows
    (iii)   ischial spines.
    (iv)   pubic tubercles.

    Objectives   12; II; 1; (1.12), (1.16).   *0.52*   37   A

776   The right ureter of the female

    (i)   opens into the base of the bladder.
    (ii)   passes through the urogenital diaphragm.
    (iii)   lies on the psoas major muscle.
    (iv)   opens into the apex of the bladder.

    Objectives   12; II; 1; (1.11), (1.12).   *0.76*   87   B
                                                  0.60   77
                                                0.63   97

777   The parts of the bladder wholly covered by
      peritoneum include the

    (i)   base.
    (ii) infero-lateral surfaces.
    (iii) neck.
    (iv) superior surface.

Objectives   12; II; 2; (2.2), (2.3).   *0.56* 92 D
                                           0.46 92

778   Structures behind the male bladder include the

    (i)   seminal vesicles.
    (ii) ampullae of the ductus deferentes.
    (iii) rectum.
    (iv) ureters.

Objectives   12; II; 2; (2.1), (2.3).   *0.29* 42 E

779   Constituents of the spermatic cord include the

    (i)   ureter.
    (ii) testicular artery.
    (iii) epididymis.
    (iv) pampiniform plexus of veins.

Objectives   12; III; 1; (1.7).   *0.55* 95 C

780   The uterus

    (i)   forms an angle of 90 degrees or more with the
        vagina.
    (ii) is covered by peritoneum above the isthmus.
    (iii) has a supravaginal part of the cervix.
    (iv) is most importantly supported by the round
        ligaments.

Objectives   12; IV; 1; (1.1), (1.3), (1.6), (1.7).   *0.35* 79 A

781   The uterus lies

    (i)   superior to the bladder.
    (ii) anterior to the rectum.
    (iii) partly within the vagina.
    (iv) vertically in the mid-line.

Objectives   12; IV; 1; (1.1), (1.5), (1.7).   *0.54*   84   A

782   The cervix

    (i)   is related laterally to the ureters and uterine
          arteries.
    (ii) is the most freely movable part of the uterus.
    (iii) is separated from the rectum by the recto-uterine
          pouch.
    (iv) drains its lymph directly to the lumbar nodes.

Objectives   12; IV; 1; (1.1), (1.5), (1.6).
                 12; V; 1; (1.6).   *0.51*   82   B

783   The pelvic peritoneum in the female

    (i)   covers the whole body of the uterus.
    (ii) does not completely cover the uterine tubes.
    (iii) is reflected from the tubes as the infundibulo-
          pelvic folds.
    (iv) covers the upper two-thirds of the front of the
          rectum.

Objectives   12; IV; 1; (1.5).   —   —   E

784   The vagina

    (i)   has an anterior wall which is longer than the
          posterior.
    (ii) lies anterior to the urethra.
    (iii) is separated from the bladder by peritoneum.
    (iv) has an anterior wall which is pierced by the
          cervix.

Objectives   12; IV; 1; (1.7).   *0.38*   59   D
                                           0.38   96

785 The vestibule

(i)   lies at the upper end of the vagina.
(ii) receives the urethral orifice.
(iii) is related to the recto-uterine pouch.
(iv) is the cleft between the labia minora.

Objectives   12; IV; 1; (1.8).   —   —   C

786 The labia minora

(i)   are usually hidden by the labia majora.
(ii) form the prepuce of the clitoris.
(iii) bound the vestibule of the vagina.
(iv) form the frenulum of the clitoris.

Objectives   12; IV; 1; (1.8).   0.43   57   E

K(iv) Blood supply, lymphatic drainage and innervation (Questions 787–799)

787 Branches of the coeliac trunk include the

(i)   right gastric artery.
(ii) left gastric artery.
(iii) right gastro-duodenal artery.
(iv) splenic artery.

Objectives   12; I; 5; (5.1).   0.25   72   C
                                                     0.49   87

788 The superior mesenteric artery is distributed to the

(i)   ascending colon.
(ii) transverse colon.
(iii) ileum.
(iv) descending colon.

Objectives   12; I; 5; (5.2).   0.56   92   A

789   Branches of the superior mesenteric artery supply the

   (i)   whole duodenum.
   (ii) right colic flexure.
   (iii) greater part of the pancreas.
   (iv) appendix.

   Objectives   12; I; 5; (5.2).   *0.65*   86   C

790   Tributaries of the portal venous system include the

   (i)   splenic vein.
   (ii) hepatic veins.
   (iii) inferior mesenteric vein.
   (iv) left gonadal vein.

   Objectives   12; I; 5; (5.4).   *0.32*   91   B

791   Principal portal-systemic venous anastomoses occur

   (i)   in the liver.
   (ii) in the anal canal.
   (iii) in the spleen.
   (iv) at the lower end of the oesophagus.

   Objectives   12; I; 5; (5.5).   *0.55*   50   C
                                                    0.36   59
                                                    0.32   67

792   Blood from the liver drains

   (i)   directly into the inferior vena cava.
   (ii) directly into the portal vein.
   (iii) via sinusoids into the portal vein.
   (iv) by hepatic veins from the right and left lobes
         of the liver.

   Objectives   12; I; 5; (5.4); 6; (6.5).   —   —   D

793   Venous drainage from the bladder may reach the

(i)   internal vertebral venous plexus.
(ii) inferior mesenteric vein.
(iii) internal iliac veins.
(iv) gonadal veins.

Objectives   12; V; 1; (1.6).   *0.34*   73   A

794   The lymphatic vessels from the testis drain into nodes

(i)   below the inguinal ligament.
(ii) along the internal iliac artery.
(iii) along the external iliac artery.
(iv) beside the aorta.

Objectives   12; V; 1; (1.6).   –   –   D

795   Lymph from the uterus may drain through nodes
along the

(i)   inferior mesenteric vessels.
(ii) internal iliac vessels.
(iii) internal pudendal vessels.
(iv) aorta.

Objectives   12; IV, 1; (1.9).   *0.29*   83   C

796   The vagal trunks

(i)   provide for the inhibition of peristaltic activity.
(ii) provide pre-ganglionic secretomotor innervation
      for gastric glands.
(iii) provide post-ganglionic motor fibres to the gastric
      musculature.
(iv) are distributed to the alimentary canal from the
      oesophagus to the left colic flexure.

Objectives   12; I; 5; (5.9).   *0.34*   55   C

797   The coeliac plexus

(i)   surrounds the coeliac trunk.
(ii) includes pre-ganglionic sympathetic fibres to the
      ileum.
(iii) includes afferent fibres from the duodenum.
(iv) includes parasympathetic fibres to the sigmoid
      (pelvic) colon.

Objectives   12; I; 5.   *0.28*   47   A

798   The pelvic part of the sympathetic trunk

(i)   lies medial to the pelvic sacral foramina.
(ii) contains pre-ganglionic fibres concerned with
      the innervation of the lower limb.
(iii) gives branches to the inferior hypogastric plexus.
(iv) contains cells of origin of post-ganglionic fibres
      to the lower limb.

Objectives   12; V; 1.   —   —   E

799   The pelvic splanchnic nerves

(i)   are branches of the sympathetic trunks.
(ii) carry motor fibres to the descending colon.
(iii) carry post-ganglionic parasympathetic fibres.
(iv) convey sensory fibres from the trigone of the
      bladder.

Objectives   12; V; 1; (1.5).   *0.47*   41   C

Chapter 8

Neurobiology

Type A items (Hubbard and Clemans, 1961) (Questions 800–879)

These involve choosing one correct answer from five available choices. The instruction for these items is as follows:
Each of the incomplete statements below is followed by five suggested answers or completions. Select the one which is best in each case.

A(i) Spinal cord (Questions 800–807)

800 The width of the spinal cord is notably increased

    (A) between cord levels C1 and C4.
    (B) between cord levels C5 and T1.
    (C) between vertebral levels C2 and C3.
    (D) where a lateral column of grey matter is found.
    (E) in none of the above situations.

Objectives   13; 1; 1; (1.1), (1.3), (1.8).   —   —   B

801 In relation to the structure of the spinal cord

    (A) its extent is from foramen magnum to L3.
    (B) lower cervical segments show the presence of a
        lateral grey column.
    (C) middle sacral segments show the presence of a
        lateral grey column.

(D) the central canal is not continuous with the
fourth ventricle.
(E) the dorsal surface is marked by the presence of
a deep median fissure.

Objectives   13; I; 1; (1.1), (1.3), (1.8).
              13; III; 1; (1.1).   *0.48* 60 C

802 The lower end of the spinal cord is found at the
level of

(A) the 2nd sacral vertebra.
(B) the 5th sacral vertebra.
(C) the 4th lumbar vertebra.
(D) the 1st lumbar vertebra.
(E) none of the above.

Objectives   13; I; 1.   *0.23* 71 D

803 The spinal dural sac usually terminates opposite

(A) the lower border of the 2nd sacral vertebra.
(B) the centre of the 4th lumbar vertebra.
(C) the lower border of the 1st lumbar vertebra.
(D) the lower border of the 3rd lumbar vertebra.
(E) none of the above situations.

Objectives   13; I; 1.   *0.23* 66 A
                                                 0.38 87
                                                 0.18 74

804 Cord tracts involved in the transmission of signals
subserving touch

(A) are all crossed.
(B) include the fasciculus cuneatus.
(C) all have cells of origin in the chief nucleus of
the posterior grey column.
(D) include the lateral spinothalamic tract.
(E) include fibres originating in the substantia
gelatinosa.

Objectives   13; I; 1; (1.3), (1.4), (1.5), (1.7).   — — B

805 The lateral spinothalamic tract

(A) conveys painful and thermal sensitivity from the opposite side of the body.
(B) conveys tactile sensibility from the opposite side of the body.
(C) conveys painful and thermal sensibility from the same side of the body.
(D) has its cells of origin in spinal ganglia.
(E) has none of the above properties.

Objectives   13; I; 1; (1.4), (1.5), (1.7).   *0.83*   96   **A**

806 The lateral corticospinal tract

(A) commences in the upper part of the medulla.
(B) contains fibres which mainly terminate in synaptic contact with the motor neurons of the anterior grey column.
(C) is composed mainly of uncrossed fibres.
(D) has fibres which terminate in synaptic connection with anterior column interneurones.
(E) has none of the above properties.

Objectives   13; I; 1; (1.1), (1.4), (1.6).
                 13; I; 2; (2.2).   *0.17*   62   **D**

807 The results of hemisection of the spinal cord include

(A) loss of ipsilateral tactile sensibility.
(B) loss of contralateral proprioceptive sensibility.
(C) loss of ipsilateral temperature sensibility.
(D) loss of ipsilateral pain sensibility.
(E) none of the above.

Objectives   13; I; 1; (1.9).   *0.58*   83   **E**

A(ii) Brain stem (Questions 808–815)

808 Sections of mid-brain at the level of the superior colliculus are notable for the

(A) medial lemniscus lying next to the mid-line.

(B) absence of corticospinal fibres.
(C) small size of the substantia nigra.
(D) emerging fibres of the oculomotor nerve.
(E) emerging fibres of cranial nerve IV.

Objectives   13; l; 1; (2.1), (2.3); 3; (3.2), (3.3).   *0.46*   95   D
                                                           0.58   83
                                                           0.78   88

809   The nucleus of the solitary tract receives fibres from
      the following cranial nerves

      (A) VI, VII, X.
      (B) VII, X, XI.
      (C) VII, IX, X.
      (D) IX, X, XI.
      (E) V, VII, X.

      Objectives   13; l; 3; (3.1), (3.3), (3.6).   *0.55*   87   C

810   The nucleus of the solitary tract

      (A) contains the cell bodies of primary afferent
          neurones.
      (B) is concerned with the sensation of taste.
      (C) extends throughout the length of the brain stem.
      (D) lies in the centre of the solitary tract.
      (E) is concerned with the sensation of touch from
          the facial skin.

      Objectives   13; l; 1; (3.3), (3.6).   *0.48*   91   B

811   The nucleus ambiguus

      (A) contains both motor and sensory neurons.
      (B) gives origin to motor fibres to the muscles of
          mastication.
      (C) lies medial and parallel to hypoglossal nucleus.
      (D) gives origin to all the motor fibres of cranial
          nerves IX, X and XI.
      (E) has none of the above properties.

      Objectives   13; l; 3; (3.3), (3.6).   *0.49*   73   E

812 The facial nerve is directly connected to the

(A) red nucleus.
(B) inferior olivary nucleus.
(C) middle cerebellar peduncle.
(D) Edinger–Westphal nucleus.
(E) nucleus of the solitary tract.

Objectives   13; l; 3; (3.6).   0.58   87   E
                                                 0.64   86

813 The trigeminal nuclear complex

(A) has subdivisions related to different sensory
    modalities.
(B) gives rise to fibres which emerge from the brain
    stem at the superior end of the medulla oblongata.
(C) is restricted to the pons and medulla oblongata.
(D) is reached only by afferents from the trigeminal
    nerve.
(E) lies medial to the general somatic efferent nuclei
    of the brain stem.

Objectives   13; l; (3.1), (3.3), (3.4), (3.6).   0.46   77   A
                                                  0.46   80
                                                  0.64   92

814 The branchial (special visceral) efferent nuclei of the
brain stem include the

(A) nucleus of the solitary tract.
(B) dorsal nucleus of the vagus.
(C) red nucleus.
(D) hypoglossal nucleus.
(E) motor nucleus of the trigeminal nerve.

Objectives   13; l; 3; (3.1), (3.4).   0.61   56   E
                                                  0.55   46
                                                  0.60   53
                                                  0.59   78

815   The crossing of fibres which form the medial lemniscus

(A) ensures that pain and temperature are bilaterally
represented in the cerebral cortex.
(B) occurs inferior to the vestibular nuclear complex.
(C) results in loss of somatotopic organisation in
this pathway.
(D) occurs in the lowermost levels of the medulla.
(E) is only partial and 15% of fibres remain
uncrossed.

Objectives   13; l; 2; (2.1), (2.2).   *0.60*   73   B
                                               0.58   74
                                               0.46   81

A(iii) Sensory system (Questions 816–820)

816   Sagittal division of the optic chiasma results in

(A) total blindness.
(B) loss of the consensual light reflex.
(C) loss of vision in both nasal fields.
(D) loss of vision in both temporal fields.
(E) total blindness but preservation of visual
reflexes.

Objectives   13; ll; 2; (2.4), (2.5).   *0.43*   73   D
                                               0.48   88
                                               0.58   88

817   The axons of the ganglion cells in one retina project

(A) wholly to the ipsilateral lateral geniculate nucleus.
(B) wholly to the contralateral lateral geniculate
nucleus.
(C) to both ipsilateral and contralateral lateral
geniculate nuclei.
(D) to the visual cortex.
(E) to the bipolar cells of the retina.

Objectives   13; ll; 2; (2.1), (2.4).   *0.43*   93   C
                                               0.38   96
                                               0.92   96

818   The lateral lemniscus

(A) arises in the olivary nucleus.
(B) carries proprioceptive sensation from the head
and neck.
(C) ends in the superior colliculus.
(D) ends in the red nucleus.
(E) has none of the above properties.

Objectives   13; II; 1; (1.3), (1.4).   0.45   81   E
                                                    0.68   86
                                                    0.61   73

819   Complete transection of the right internal capsule
results in

(A) anosmia on the right side (right loss of olfaction).
(B) ipsilateral paralysis and somatosensory loss.
(C) left homonymous hemianopia.
(D) contralateral loss of pain and temperature
without loss of touch sensation.
(E) loss of hearing in the right ear.

Objectives   13; I; 6; (6.4).   0.47   37   C
                                                    0.60   44
                                                    0.53   75

820   Nystagmus may result from

(A) damage to the medial geniculate body.
(B) cutting one optic nerve.
(C) damage to the medial lemniscus.
(D) damage to the cerebellum.
(E) none of the above.

Objectives   13; II; 2; (2.8).   —   —   D

A(iv) Motor system and cerebellum (Questions 821–827)

821   Unilateral destruction of the motor cortex
(pre-central gyrus and pre-motor area) causes

(A) contralateral paresis and hypotonia.
(B) contralateral paresis and hypertonia.

(C) contralateral incoordination and hypertonia.
(D) ipsilateral paresis and some sensory deficit.
(E) contralateral tremor and hypertonia.

Objectives   13; II; 3; (3.1), (3.4), (3.5).   *0.52*   73   B
                                                      0.35   59

822   The decussation of the pyramids

(A) is found at the level of the 4th ventricle.
(B) is not uncommonly absent.
(C) represents the crossing of main sensory tracts.
(D) occurs rostral to the 'sensory' decussation.
(E) is usually incomplete.

Objectives   13; I; 1; (1.6); 2; (2.1), (2.2).   *0.51*   74   E
                                                      0.57   80

823   The lateral corticospinal tract

(A) contains fibres principally derived from cells in
     the parietal cortex.
(B) contains fibres which mainly terminate in
     synaptic contact with motor neurones.
(C) is composed of crossed fibres.
(D) commences in the pons.
(E) terminates distally in the mid-thoracic region of
     the spinal cord.

Objectives   13; I; 1; (1.6); 2; (2.2).   13; II; 3; (3.1).   *0.40*   73   C
                                                             0.51   79
                                                             0.45   90

824   The cerebellar cortex has *direct* connections with

(A) the vestibular nuclei.
(B) the cerebral cortex.
(C) anterior horn cells of the spinal cord.
(D) the substantia nigra.
(E) the nucleus ambiguus.

Objectives   13; I; 5; (5.3).   *0.38*   71   A
                                                      0.47   78
                                                      0.57   91

825   A large lesion of the left cerebellar hemisphere
      would result in

   (A) loss of conscious proprioception from the
       right side of the body.
   (B) loss of conscious proprioception from the left
       side of the body.
   (C) incoordination of voluntary movement of the
       left limbs.
   (D) exaggerated tendon reflexes on the left side of
       the body.
   (E) none of the above.

   Objectives   13; l; 5; (5.3).   *0.39* 48   C
                                                    0.40 60

826   The dentate nucleus sends fibres to the

   (A) globus pallidus.
   (B) substantia nigra.
   (C) spinal cord.
   (D) thalamus.
   (E) pons.

   Objectives   13; l; 5; (5.1), (5.3).   *0.69* 89   D

827   The tone and posture of the limbs is directly
      influenced by the

   (A) brain stem reticular formation.
   (B) hypothalamohypophysial tract.
   (C) dorsal nucleus of the vagus.
   (D) nucleus ambiguus.
   (E) trapezoid body.

   Objectives   13; l; 4; (4.2).   13; ll; 4; (4.2).   *0.42* 87   A
                                                       0.48 93
                                                       0.87 96

A(v) The circulation of the brain, and the cerebrospinal fluid (Questions 828–831)

828 The cortical branches of the middle cerebral artery supply

(A) all of the precentral gyrus.
(B) all of the primary auditory cortex.
(C) a small part of the primary visual cortex.
(D) the cingulate gyrus.
(E) all of the primary somatosensory cortex.

Objectives   13; III; 3; (3.4).   *0.45   72   B*
                                                   0.56   76

829 The middle meningeal vessels run

(A) in the subdural plane.
(B) in the subarachnoid plane in the floor of the middle cranial fossa and then in the extradural plane.
(C) in the extradural plane.
(D) together but with the artery extradural and the vein subdural.
(E) in the depth of the scalp.

Objectives   13; III; 1.   *0.41   86   C*

830 The straight sinus

(A) is formed by the union of the inferior sagittal sinus and the superior petrosal sinus.
(B) runs in the free lower border of the falx cerebri.
(C) joins the great cerebral vein to form the inferior sagittal sinus.
(D) runs in the attached margin of the tentorium cerebelli.
(E) has none of the above properties.

Objectives   13; III; 3; (3.7), (3.8).   *0.30   25   E*
                                                   0.45   45

831   The superior sagittal sinus

(A) passes through the jugular foramen.
(B) joins the inferior sagittal sinus.
(C) traverses the foramen caecum.
(D) joins the transverse sinus.
(E) has none of the above properties.

Objectives   13; III; 3; (3.8).   *0.15*   85   D
                                                          0.25   84
                                                          0.36   89

A(vi) Cerebrum (Questions 832–839)

832   Motor speech areas of the cerebral cortex are located

(A) bilaterally in the precentral gyrus.
(B) in the precentral gyrus of the dominant
    hemisphere.
(C) bilaterally in the inferior frontal gyrus.
(D) in the inferior frontal gyrus of the dominant
    hemisphere.
(E) in the occipital lobe.

Objectives   13; III; 2; (2.1), (2.3).   *0.20*   66   D

833   In the cerebral cortex of a right-handed person,
      speech areas are usually located

(A) in the inferior temporal gyrus of the temporal
    lobe.
(B) bilaterally in the inferior frontal gyrus.
(C) bilaterally in the precentral gyrus.
(D) in the inferior frontal gyrus of the left
    hemisphere.
(E) in the precentral gyrus of the left hemisphere.

Objectives   13; I; 6; (6.6).   13; III; 2; (2.1), (2.3).   *0.55*   87   D

834   A sagittal surgical division of the whole of the corpus
      callosum would result in

(A) incoordination between right and left limbs.
(B) loss of vision in both temporal fields.

(C) severe psychiatric disturbance due to splitting of the personality.
(D) loss of most but not all of the forebrain commissural fibres.
(E) loss of orientation of the body in space.

Objectives   13;1;6;(6.6).   *0.39*   72   D
                                              0.40   76
                                              0.48   86

835   The central sulcus separates the

(A) area of termination of the spinothalamic system from that of the dorsal columns.
(B) motor cortex from the auditory cortex.
(C) motor cortex from the area of termination of the somatosensory system.
(D) primary somatosensory area from the somatosensory association area.
(E) area of cortical supply of the anterior cerebral artery from that of the middle cerebral artery.

Objectives   13;1;6;(6.1).   13;III;3;(3.4).   *0.55*   97   C
                                               1.06   99
                                               0.61   97

836   The internal capsule is made up of

(A) projection and association fibres.
(B) commissural and cortical efferent fibres.
(C) association fibres only.
(D) cortical efferent fibres only.
(E) cortical efferent and cortical afferent fibres.

Objectives   13;1;6;(6.1),(6.3).   *0.42*   90   E

837   The internal capsule

(A) lies between the thalamus and the caudate nucleus.
(B) is principally composed of corticospinal fibres.
(C) lies between the putamen and the globus pallidus.

    (D) carries fibres of the optic but not the auditory
        radiation.
    (E) carries thalamocortical projections.

    Objectives   13; I; 6; (6.1), (6.2), (6.3).   *0.49*   50   E
                                                           0.62   87

838 The mamillary bodies

    (A) lie rostral to the tuber cinereum.
    (B) receive input from the fornix.
    (C) project fibres to the hypophysis cerebri.
    (D) project fibres to the olivary nucleus.
    (E) lie caudal to the posterior perforated substance.

    Objectives   13; III; 4; (4.2).   *0.45*   61   B

839 The fornix

    (A) is the main projection pathway of the
        hippocampus.
    (B) lies wholly posterior to the thalamus.
    (C) terminates in the thalamus.
    (D) forms part of the roof of the lateral ventricle.
    (E) lies posterior to the interventricular foramen.

    Objectives   13; I; 6; (6.2).   13; III; 4; (4.2).   *0.36*   83   A

A(vii) Physiology (Questions 840–878)

840 The inhibitory post-synaptic potential in a mammalian
     motor neurone produced by a single shock to Ia nerve
     fibres is

    (A) due to the release of glutamate by the Ia fibre
        on to the motor neurone.
    (B) due to the activity of an interneurone releasing
        gamma-aminobutyric acid.
    (C) due to the release of glycine by the Ia fibre.
    (D) a hyperpolarization of the motor neurone
        membrane lasting about 150 msec.

(E) due to an increased permeability of the motor
neurone membrane to potassium and chloride
ions.

Objectives —   *0.39*   84   E

841 Ia afferent fibres from a muscle

(A) directly excite gamma motor neurones.
(B) directly inhibit motor neurones going to
antagonistic muscles.
(C) directly excite motor neurones going to
synergistic muscles.
(D) indirectly inhibit motor neurones going to
synergistic muscles.
(E) directly excite inhibitory interneurones going
to synergistic muscles.

Objectives —   *0.43*   83   C

842 Gamma motor neurones

(A) are present only in nerves to postural muscles.
(B) have cell bodies in the lateral horn of the spinal
cord.
(C) synapse with Ia afferent fibres.
(D) have conduction velocities in the range
20—40 m/s.
(E) are co-activated with alpha motor neurones in
the phasic stretch reflex.

Objectives   13; II; 3; (3.3).   *0.42*   41   D

843 The Purkinje cells of the cerebellar cortex

(A) receive an input from mossy fibres.
(B) are innervated mainly by climbing fibres.
(C) are inhibited by granule cells.
(D) project to the deep cerebellar nuclei.
(E) are excitatory to cells of the dentate nucleus.

Objectives   13; I; 5; (5.4). 13; II; 3; (3.1).   *0.66*   89   D

844   The body-on-head righting reflex

   (A) is initiated in the labyrinth.
   (B) is abolished by section of the first four cervical
       dorsal roots.
   (C) is initiated by flexion of the neck.
   (D) can occur in a mid-collicular decerebrate animal.
   (E) is reduced or abolished by placing a weighted
       board on the animal.

   Objectives   —   *0.31*   34   E

845   The delta rhythm

   (A) has a frequency in the range 4–8 Hz.
   (B) is absent in babies.
   (C) is the predominant rhythm of slow-wave sleep.
   (D) occurs during episodes of frustration.
   (E) is best recorded over the occipital scalp.

   Objectives   —   *0.33*   69   C

846   In the auditory system, tuning curves are

   (A) a plot of firing frequency as a function of
       intensity.
   (B) a plot of threshold as a function of frequency.
   (C) broadened by lateral inhibition.
   (D) narrower in VIII nerve neurones than in medial
       geniculate body.
   (E) obtained only in basilar membrane.

   Objectives   —   *0.41*   90   B

847   The 'satiety centre' is situated in the

   (A) lateral hypothalamic nucleus.
   (B) locus coeruleus.
   (C) mamillary bodies.

(D) supra-optic nucleus of the hypothalamus.
(E) ventromedial nucleus of the hypothalamus.

Objectives   13; III; 4; (4.3).   *0.47*   89   E

848 A generator potential

(A) does not adapt during prolonged stimulation.
(B) is graded.
(C) is not spatially summed.
(D) is propagated.
(E) has none of the above properties.

Objectives   –   *0.43*   89   B

849 All bipolar cells in the vertebrate retina

(A) respond to an increase in the general level of
    illumination.
(B) generate action potentials of constant amplitude
    when the excitation reaches threshold.
(C) have receptive fields with an antagonistic
    centre-surround organization.
(D) respond exclusively to a decrease in the level of
    illumination in the centre of their receptive fields.
(E) interact with each other through direct
    reciprocal synaptic contacts.

Objectives   –   *0.56*   90   C

850 The excitatory post-synaptic potential (EPSP) in a
     mammalian motor neurone produced by stimulation
     of Ia nerve fibres

(A) is due to the release of acetylcholine by the
    afferent fibre.
(B) is a depolarization lasting about 20 ms.
(C) is due to a specific increase in the sodium
    permeability of the motor neurone membrane.
(D) has a refractory period of about 1.5 ms.
(E) propagates along dendrites at about 1 m/s.

Objectives   –   *0.36*   48   B

851 Astrocytes

(A) have low membrane resting potentials.
(B) are hyperpolarized during increased nerve activity.
(C) have membrane potentials very sensitive to changes in extracellular potassium.
(D) are non-phagocytic.
(E) form the blood—brain barrier.

Objectives   —   *0.38* 82   C

852 The Golgi tendon organ

(A) is innervated by Group II nerve fibres.
(B) is a sense organ concerned with the adjustment of tension in a muscle.
(C) mediates the 'tendon jerk'.
(D) lies in parallel with extrafusal muscle fibres.
(E) is excited by stretch of a muscle but not by its contraction.

Objectives   13; II; 4; (4.2).   *0.33* 94   B

853 The Purkinje cells of the cerebellar cortex

(A) project to the deep cerebellar nuclei.
(B) receive an input from mossy fibres.
(C) are innervated mainly by climbing fibres.
(D) can be both excitatory and inhibitory to other cells.
(E) project to the thalamus.

Objectives   13; I; 5; (5.4).   13; II; 3; (3.1). *0.62* 86   A

854 The alpha rhythm

(A) has a frequency of 14—60 Hz.
(B) is believed to originate in the thalamus.
(C) is present in babies.
(D) is absent in most people when the eyes are closed.

(E) is present during mental concentration.

Objectives   13; II; 2.   *0.48*   78   B

855   Retinal bipolar cells

(A) are neurones located in the inner nuclear layer
      of the retina and their axons form optic nerve
      fibres.
(B) when stimulated by light generate all-or-nothing
      action potentials.
(C) have receptive fields with an antagonistic centre-
      surround organization.
(D) are post-synaptic to the retinal ganglion cells.
(E) generate graded S-potentials.

Objectives   −   *0.54*   80   C

856   Simple cells in the visual cortex

(A) are usually located in the upper and lower
      cortical layers of the parastriate area.
(B) are orientation-selective and in the majority of
      cases can be activated through either eye.
(C) respond optimally to very-fast-moving objects.
(D) receive their principal excitatory input from
      complex cells.
(E) have receptive fields which are usually larger
      than the receptive fields of complex cells.

Objectives   −   *0.51*   87   B

857   Light shone in the left eye produced a pupillary
      constriction only in the right eye; light shone in
      the right eye produced pupillary constriction only
      in the right eye. The lesion would have been in the

(A) left oculomotor nerve.
(B) left optic nerve.

(C) left pretectum.
(D) right oculomotor nerve.
(E) right optic nerve.

Objectives   13; II; 2; (2.6).   *0.38*   88   A

858   A 5D myope has a range of accommodative power
       of 10D. His far point will therefore be at

(A) 0.01 m.
(B) 0.1 m.
(C) 0.2 m.
(D) 0.5 m.
(E) 5 m.

Objectives   —   *0.41*   75   C

859   When environmental temperatures are very high,
       heat loss is achieved by

(A) activation of parasympathetic systems.
(B) increasing convective loss.
(C) increasing conductive loss.
(D) metabolic inhibition.
(E) vaporization.

Objectives   —   *0.44*   91   E

860   Post-synaptic inhibition depends upon

(A) a specific increase of membrane permeability to
     potassium ions.
(B) a hyperpolarization of the cell that is inhibited.
(C) a specific increase of membrane permeability to
     chloride ions.
(D) an increase of membrane permeability to species
     of ions with an equilibrium potential below the
     threshold for excitation.
(E) a depolarization of excitatory terminals on the
     cell which is inhibited.

Objectives   —   *0.24*   46   D

861 Normally an action potential in a motor neurone would be initiated in the

(A) initial segment.
(B) soma.
(C) soma-dendritic region.
(D) large dendrites.
(E) recurrent collateral branch of the axon.

Objectives —   *0.50* 91   A

862 When a peripheral nerve is made anoxic, the sensation remaining longest is

(A) cold.
(B) pain.
(C) proprioception.
(D) touch/pressure.
(E) warmth.

Objectives —   *0.26* 59   B

863 A myope has a near point of 5 cm and a far point of 50 cm. The amount of his myopia is

(A) 0.1 D.
(B) 0.2 D.
(C) 0.5 D.
(D) 2 D.
(E) 10 D.

Objectives —   *0.53* 74   D

864 Lateral head tilt is an adequate stimulus to hair cells situated in the

(A) ampulla.
(B) crista of the horizontal semicircular canal.
(C) crista of the superior semicircular canal.
(D) crista of the posterior semicircular canal.
(E) macula of the utricle.

Objectives —   *0.67* 73   E

865   Characteristics of the fovea of the human eye include

(A) cones of only two types.
(B) considerable convergence of cones on to bipolar cells.
(C) decreased density of retinal (pigment) epithelium.
(D) more numerous bipolar and ganglion cells.
(E) shorter cones than in peripheral retina.

Objectives   13; II; 2; (2.2).   0.33   46   A

866   The c wave of the electroretinogram is due to

(A) bipolar cells.
(B) cones.
(C) glial cells.
(D) retinal (pigment) epithelium.
(E) rods.

Objectives   —   0.37   75   D

867   Visual cells which do not have spatially separate 'on' and 'off' areas in their receptive fields are

(A) amacrine cells in the retina.
(B) bipolar cells in the retina.
(C) complex cells in the visual cortex.
(D) simple cells in the visual cortex.
(E) X-type retinal ganglion cells.

Objectives   —   0.24   35   C

868   Lesions in the superior colliculus affect

(A) auditory discrimination.
(B) auditory localization.
(C) balance.
(D) visual discrimination.
(E) visual orientation.

Objectives   13; II; 2; (2.6).   0.49   76   E

869   Properties common to all rapidly adapting generator
      potentials of mechanoreceptors include

      (A) gradation with intensity of stimulus.
      (B) location in skin.
      (C) 'off' response.
      (D) transient and sustained components in response.
      (E) none of the above.

      Objectives   –   *0.35*   53   A

870   Cells in the cuneate nucleus

      (A) may be inhibited presynaptically.
      (B) give rise to the cuneate tract.
      (C) respond well to nociceptive stimuli.
      (D) send a major projection to the ventrolateral
          nucleus of the thalamus.
      (E) send a major projection to the cerebellum.

      Objectives   13; II; 1.   *0.44*   41   A

871   It was noted in a patient that light shone in the left
      eye elicited pupillary constriction in both eyes. Light
      shone in the right eye failed to produce pupillary
      constriction in either eye. The lesion must have been
      in the

      (A) left pretectum.
      (B) right lateral geniculate nucleus.
      (C) right oculomotor nerve.
      (D) right optic nerve.
      (E) right pretectum.

      Objectives   13; II; 2; (2.6).   *0.31*   66   D

872   When testing a patient's hearing, it was found that
      the energy required for him to hear tones of 1000 Hz
      in his right ear was 1000 times that for his left ear.
      This then represents an example of

      (A) a difference of 33 dB between the two ears.
      (B) a difference of 30 dB between the two ears.

(C) a difference of 100 dB between the two ears.
(D) conduction deafness.
(E) high-frequency hearing loss.

Objectives   —   *0.45*   75   **B**

873   A particular shade of green may not be distinguished
        from grey by a protanope because

(A) his abnormal pigment absorbs equally from
      background and symbol.
(B) the photon catch for the symbol differs from
      the background only for green cones.
(C) the photon catch for the symbol differs from
      the background only for red cones.
(D) the symbol activates only green cones.
(E) the symbol does not activate red cones.

Objectives   —   *0.45*   40   **C**

874   Astrocytes

(A) have low membrane resting potentials.
(B) develop long-lasting action potentials.
(C) are responsible for the myelin sheaths of
      neurones in the central nervous system.
(D) have membrane potentials very sensitive
      to changes in extracellular potassium.
(E) form the blood-brain barrier.

Objectives   —   *0.33*   83   **D**

875   Gamma motor neurones

(A) supply the intrafusal muscle fibres.
(B) supply the extrafusal muscle fibres.
(C) supply slow (red) muscle fibres.
(D) have cell bodies in the lateral horn of the spinal
      cord.
(E) synapse with Ia afferent fibres.

Objectives   13; II, 3; (3.3).   *0.59*   91   **A**

876 The Purkinje cells of the cerebellar cortex

   (A) project to the deep cerebellar nuclei.
   (B) have glycine as their synaptic transmitter.
   (C) are activated mainly by climbing fibres.
   (D) can be both excitatory and inhibitory to other
       cells.
   (E) are the most numerous cells in the cerebellar
       cortex.

Objectives   13;1;5;(5.4).   13;II;3;(3.1).   0.47   73   A

877 The basal ganglia

   (A) mediate slow pain sensation.
   (B) lack direct connections with cerebral cortex.
   (C) are mainly connected with cerebellar cortex.
   (D) cause facilitation of stretch reflexes when
       excited.
   (E) are concerned with righting reflexes.

Objectives   13;II;3;(3.1).   0.29   32   E

878 The alpha rhythm

   (A) is best recorded from the frontal scalp.
   (B) has a frequency of 14–60 Hz.
   (C) is absent in most people when the eyes are open.
   (D) is a sign of incipient epilepsy.
   (E) is enhanced in coma.

Objectives   −   0.30   86   C

Type E items (Hubbard and Clemans, 1961) (Questions 879–937)

These items are of the assertion–reason type and there are five possible
responses. The instruction for these items is as follows:
   *Each question consists of an assertion and a reason. Responses should be
   chosen as follows:*

(A) if the assertion and reason are true statements and the reason is a correct explanation of the assertion;
(B) if the assertion and reason are true statements but the reason is NOT a correct explanation of the assertion;
(C) if the assertion is true but the reason is a false statement;
(D) if the assertion is false but the reason is a true statement;
(E) if both assertion and reason are false statements.

E(i) Spinal cord (Questions 879–882)

879   Grey rami communicantes are restricted to the thoracolumbar region of the spinal cord

BECAUSE

pre-ganglionic sympathetic fibres emerge only from the thoracolumbar regions of the spinal cord.

Objectives 13; III; 5; (5.2).	*0.32* 48	D
	0.42 67	

880   Damage to the fasciculus cuneatus does not necessarily impair motor control

BECAUSE

the fasciculus cuneatus is an ascending rather than a descending tract.

Objectives 13; I; 1; (1.4), (1.5), (1.9), (1.12).	*0.30* 34	D

881   Destruction of the dorsal columns does not result in complete loss of touch sensation below the lesion

BECAUSE

the anterior spinothalamic tract also carries touch sensation.

Objectives 13; I; 1; (1.4), (1.5), (1.7), (1.9).	*0.20* 89	A
	0.26 88	
	0.27 91	

882   Cutting the lateral spinothalamic tract at thoracic
      levels produces some sensory loss in the ipsilateral
      lower limb

      BECAUSE

      the fibres of the spinothalamic tract cross in the
      medulla oblongata.

      Objectives   13; I; 1; (1.4), (1.5), (1.10).   *0.52*   55   E
                                                             0.68   64
                                                             0.56   77

E(ii) Brain stem (Questions 883–889)

883   The nucleus of the abducent nerve is found near the
      mid-line of the brain stem

      BECAUSE

      it belongs to the general somatic efferent column of
      nuclei.

      Objectives   13; I; 3; (3.1), (3.3).   –   –   A

884   The red nucleus is easily recognized in sections at the
      level of the inferior colliculus

      BECAUSE

      the nucleus is traversed and encapsulated by fibres
      of the superior cerebellar peduncle.

      Objectives   13; I; 2; (2.1), (2.3).   *0.53*   69   D
                                                             0.54   68
                                                             0.27   51

885   The nucleus ambiguus gives rise to efferent fibres
      that travel in cranial nerves IX, X and XI

      BECAUSE

      it is the only branchial (special visceral) efferent
      nucleus in the brain stem.

      Objectives   13; I; 3; (3.1), (3.2), (3.3).   *0.53*   73   C
                                                             0.39   84

886 Damage to the nucleus ambiguus causes difficulty in swallowing

BECAUSE

its fibres are distributed to tongue musculature.

Objectives   13;1;3;(3.1),(3.2),(3.3),(3.4).   *0.51*   80   C

887 The brain stem reticular formation is restricted to the medulla oblongata

BECAUSE

the inferior olive is found in the medulla oblongata.

Objectives   13;1;2;(2.1),(2.3);4;(4.1).   *0.42*   90   D
                                                         0.53   90

888 Severe damage to the medulla oblongata is often rapidly fatal

BECAUSE

respiratory and cardiovascular centres are found in the medullary reticular formation.

Objectives   13;1;4;(4.3).   *0.41*   81   A

889 Extensive lesions of the medulla oblongata produce impairment of speech

BECAUSE

the muscles of the soft palate are supplied by the glossopharyngeal nerve.

Objectives   13;1;3;(3.2),(3.4).   −   −   C

E(iii) Sensory system (Questions 890–893)

890 Lesions of the central part of the optic chiasma may produce loss of the medial halves of the visual fields

BECAUSE

fibres from the nasal halves of both retinae cross in this situation.

Objectives 13; II; 2; (2.3), (2.5). *0.68* 93 D

891 Extirpation of the occipital lobe cortex results in loss of the pupillary light reflex

BECAUSE

the reflex pathway includes corticocollicular projections.

Objectives 13; II; 2; (2.6). — — E

892 Visual disturbances result from blockage of the cortical branches of the middle cerebral artery

BECAUSE

the middle cerebral artery supplies the cortex adjacent to the calcarine fissure.

Objectives 13; III; 3; (3.4). *0.47* 68 E
0.39 84
0.33 77

893 Extirpation of the temporal lobe cortex produces deafness

BECAUSE

the auditosensory cortex lies in the superior temporal gyrus.

Objectives 13; II; 1; (1.3), (1.4). — — D

E(iv) Motor systems and cerebellum (Questions 894—905)

894   The cerebellum is separated by layers of dura from
the occipital lobe of the cerebrum

BECAUSE

the tentorium cerebelli covers the superior surface of
the cerebellum.

Objectives   13; III; 1; (1.8).   *0.43*   64   A

895   Damage to one cerebellar hemisphere results in
motor disorders of the opposite side of the body

BECAUSE

the spinocerebellar tracts are composed of fibres all
of which cross in the spinal cord.

Objectives   13; I; 5; (5.1), (5.3).   13; II; 3; (3.1).   *0.56*   40   E
                                                             0.47   50
                                                             0.55   59

896   Lesions of the cerebellum may produce nystagmus
BECAUSE

the cerebellum is connected with the vestibular nuclei.

Objectives   13; I; 5; (5.3).   13; II; 2; (2.8).   *0.19*   46   A
                                                             0.47   82

897   Lesions of the cerebellum may produce deafness
BECAUSE

the cerebellum is directly connected with the
cochlear nuclei.

Objectives   13; I; 5; (5.1), (5.3).   13; II; 1; (1.3).   *0.42*   88   E

898   Destruction of the corticocobulbar fibres to the
facial nucleus results in incomplete facial paresis

BECAUSE

the part of the nucleus concerned with innervation
of upper facial muscles comes under bilateral cortical
control.

Objectives   13; I; 3; (3.3), (3.6).
13; II, 3; (3.1), (3.4).   *0.21*   70   A

899   The integrity of the internal capsule is vital to normal
motor function

BECAUSE

all of the major corticofugal projection fibres travel
in it.

Objectives   13; I; 6; (6.3), (6.4).   *0.17*   59   A
0.26   49

900   A complete lesion of the posterior limb of the
internal capsule would cause hemianaesthesia

BECAUSE

the fibres of the medial lemniscus traverse this
part of the capsule.

Objectives   13; I; 6; (6.3), (6.4).   13; II; 1; (1.1).   *0.34*   41   C

901   Damage to the substantia nigra may produce
important defects of motor control

BECAUSE

it projects directly to the cerebellum.

Objectives   13; II; 3; (3.1), (3.7).   *0.25*   73   C
0.54   85

902   Destruction of the cerebral hemisphere leads to
virtually total destruction of fibres in the lateral
corticospinal tract of the opposite side

BECAUSE

all of the fibres of the pyramids decussate in the
medulla oblongata.

Objectives   13; II, 3; (3.1).   0.23   88   C

903   Paresis rarely follows damage to the precentral
gyrus

BECAUSE

there are other motor pathways to the cord besides
the corticospinal tracts.

Objectives   13; II; 3; (3.1) (3.2), (3.4), (3.5).   –   –   D

904   It can be assumed that the basal ganglia play an
important part in the control of many movements

BECAUSE

most of their efferents project caudally to motor
centres in the brain stem.

Objectives   13; II; 3; (3.1), (3.7).   0.39   55   C
                                                                     0.35   44
                                                                     0.69   89

905   Rupture of the anterior division of the middle
meningeal artery can result in contralateral
paresis

BECAUSE

the middle meningeal artery supplies parts of the
motor cortex.

Objectives   13; II; 3; (3.1).   13; III; 1; (1.11).   0.31   44   C
                                                                     0.45   70
                                                                     0.42   48

E(v) The circulation of the brain and cerebrospinal fluid (Questions 906–908)

906 The cerebrospinal fluid circulates in the subdural space

BECAUSE

it is secreted into the space by the arachnoid granulations.

Objectives 13; III; 1; (1.7), (1.8), (1.9).   *0.39*   66   E

907 Blockage of the central canal of the medulla oblongata stops the secretion of cerebrospinal fluid

BECAUSE

back pressure so produced renders the choroid plexuses ineffective in secretion.

Objectives 13; III; 1; (1.1), (1.2), (1.9).   *0.50*   91   E

908 Blockage of the cortical branches of the anterior cerebral artery causes paresis in the contralateral lower limb

BECAUSE

the cortical branches of the anterior cerebral artery supply the superior part of the excitable cortex and its extension on to the medial cerebral surface.

Objectives 13; III; 3; (3.1), (3.4).   *0.33*   73   A
                                               0.57   87
                                               0.47   80

E(vi) Cerebrum (Questions 909–910)

909 The association areas of the temporal lobe of the left hemisphere of a right-handed person are important for normal speech

BECAUSE

the main motor speech area is found in the temporal lobe.

Objectives 13; III; 2; (2.2), (2.3).   *0.55*   82   C

910   The association areas of the parietal lobe of the
      left cerebral hemisphere of a right-handed person,
      are important for normal speech

      BECAUSE

      the motor speech area (Broca's area) is found here.

      Objectives   13; III; 2; (2.2), (2.3).   —   —   C

E(vii) Physiology (Questions 911–937)

911   The Golgi tendon organ is concerned with the
      regulation of tension in a muscle

      BECAUSE

      the nerves supplying the organ are excited by stretch
      of the muscle.

      Objectives   13; II; 4; (4.2).   *0.71*   19   B

912   Muscle spindles do not contribute to conscious
      proprioception

      BECAUSE

      muscle spindle information projects to the cerebellum.

      Objectives   13; II; 1; (1.1).   *0.21*   58   D

913   Dark-adaptation of the retinal rods can occur while
      wearing goggles which transmit red (long wavelength)
      light

      BECAUSE

      rods are insensitive to red (long wavelength) light.

      Objectives   —   *0.24*   80   A

914 Contraction of the stapedius muscle protects the ear from sudden loud noises

BECAUSE

contraction of the stapedius muscle reduces the mismatch of impedance between air and cochlear fluid.

Objectives — 0.42 61 E

915 The shoulder is the upper limb joint most sensitive to angular displacement

BECAUSE

mechanoreceptive neurones from the shoulder joint first synapse in the cuneate nucleus in the path to the cerebral cortex.

Objectives 13; II; 1; (1.1). 0.22 53 B

916 As rotation of the head in the horizontal plane commences, the quick phase of nystagmus is in the direction of rotation

BECAUSE

hair cells in the macula of the utricle are strongly activated by rotation of the head.

Objectives — 0.52 73 C

917 High-frequency sounds cause maximal vibration of the basilar membrane at the apex of the cochlea

BECAUSE

the basilar membrane is less compliant at the apex of the cochlea.

Objectives — 0.34 72 E

918   In a normal audiometric curve, thresholds for hearing
       are lowest for sounds of about 500 Hz

       BECAUSE

       thresholds for hearing are frequency-dependent.

       Objectives   —   *0.33*   65   D

919   Voluntary movement in the fingers is not initiated
       by activity in the gamma motor neurones

       BECAUSE

       activity in the Ia fibres does not precede the
       contraction.

       Objectives   13; II; 3; (3.3).   *0.45*   42   A

920   The crossed extensor reflex is a postural reflex

       BECAUSE

       it is associated with a flexion reflex.

       Objectives   13; II; 4; (4.2), (4.3).   *0.23*   56   B

921   Short-term memory can not be due to circulating
       nerve impulses

       BECAUSE

       circulating nerve impulses are interrupted during
       sleep, coma and anaesthesia.

       Objectives   —   *0.21*   64   D

922   An animal with a lesion of the ventromedial nucleus
       of the hypothalamus will not eat

       BECAUSE

       the hunger centre is located in the lateral
       hypothalamic area.

       Objectives   13; III; 4; (4.4).   *0.39*   56   D

923 Vibration is coded by rapidly adapting
mechanoreceptors

BECAUSE

all rapidly adapting mechanoreceptors have on and
off responses.

Objectives   −   *0.34*   82   C

924 A lesion of the spinal cord at the level of S1 may
abolish the ankle jerk

BECAUSE

the synapse for the ankle jerk reflex is situated at
the S1 level.

Objectives   13; I; 1.   *0.30*   58   A

925 Astrocytes are not concerned with neuronal function

BECAUSE

they do not give action potentials.

Objectives   −   *0.24*   73   D

926 Tendon jerks are phasic stretch reflexes

BECAUSE

they operate via a monosynaptic reflex arc.

Objectives   13; II; 4; (4.2).   *0.11*   42   B

927 The Golgi tendon organ is concerned with the
regulation of the length of skeletal muscle

BECAUSE

it is in parallel with the extrafusal fibres.

Objectives   13; II; 4; (4.2).   *0.54*   56   E

928 After horizontal rotation to the right is suddenly
stopped there is nystagmus to the right

BECAUSE

sudden stopping of rotation causes a deflection of
the cupulae of the horizontal semicircular canals
towards the utricle.

Objectives −   *0.51*   39   E

929 Hair cells near the apex of the basilar membrane in
the cochlea respond best to sounds of low frequency

BECAUSE

the auditory system is tonotopically organized.

Objectives −   *0.35*   49   B

930 When fixating a point in space, the blind spot lies in
the temporal visual field

BECAUSE

the optic disc lies on the temporal side of the blind
spot in the retina.

Objectives   13; II; 2; (2.2).   *0.25*   72   C

931 The 'gate control' theory predicts that removal of
or damage to large fibres in peripheral nerve will
reduce pain sensations

BECAUSE

large and small fibre systems interact in the spinal
cord.

Objectives −   *0.60*   75   D

932 A hypermetropia is corrected by a biconvex lens

BECAUSE

the refracting system is insufficiently powerful
for the length of the eyeball.

Objectives −   *0.28*   72   A

933 The metacarpophalangeal joint is more sensitive to angular displacement than the shoulder

BECAUSE

the innervation of distal joints is denser than the innervation of proximal joints.

Objectives —   *0.35*   39   E

934 The ossicles in the middle ear help to reduce the impedance mismatch between air and cochlear fluid

BECAUSE

the basilar membrane is narrower and stiffer at its base near the oval window.

Objectives —   *0.24*   77   B

935 It is possible to describe four modalities of taste

BECAUSE

taste receptors respond uniquely to one modality.

Objectives —   *0.40*   71   C

936 When environmental temperatures are very high, vaporization is the only method of heat loss available

BECAUSE

the anterior hypothalamus initiates sweating when environmental temperatures are very high.

Objectives —   *0.36*   50   B

937 The Golgi tendon organ is concerned with the regulation of length in a muscle

BECAUSE

it is in series with the extrafusal muscle fibres.

Objectives   13; II; 4; (4.2).   —   —   D

Type K items (Hubbard and Clemans, 1961) (Questions 938–1000)

These items are in the form of incomplete statements and call for recognition of one or more correct responses which are grouped in various ways. The instruction for these items is as follows:

For each of the incomplete statements below, one or more of the completions given is correct. Responses should be chosen as follows:

(A) if only completions (i), (ii) and (iii) are correct,
(B) if only completions (i) and (iii) are correct,
(C) if only completions (ii) and (iv) are correct,
(D) if only completion (iv) is correct,
(E) if all completions are correct.

K(i) Spinal cord (Questions 938–941)

938   Thoracic levels of spinal cord are characterized by

    (i)   relatively thin dorsal and ventral columns of grey matter.
    (ii) the presence of a lateral grey column.
    (iii) the presence of preganglionic autonomic neurons.
    (iv) the nucleus dorsalis which gives origin to the lateral spinothalamic tract.

Objectives	13; I; 1; (1.1), (1.3), (1.4), (1.8).	0.47	83	A
	13; III; 5; (5.2).	0.29	67	
		0.56	82	

939   The fibres of the posterior spinocerebellar tract

    (i)   begin in upper thoracic segments.
    (ii) transmit signals from muscles of the lower limbs to the cerebellum.
    (iii) arise from spinal ganglion cells of the same side.
    (iv) arise from cells of the nucleus dorsalis of the same side.

Objectives	13; I; 1; (1.3), (1.4), (1.7).	–	–	C

940 Tracts found in the lateral funiculus of the spinal
cord include the

(i) fasciculus gracilis.
(ii) rubrospinal tract.
(iii) anterior corticospinal tract.
(iv) anterior spinocerebellar tract.

Objectives 13; I; 1; (1.4).   *0.42*   90   C
                                          0.57   91

941 Sensory modalities subserved by the fasciculus
cuneatus include

(i) touch.
(ii) pain.
(iii) position sense.
(iv) temperature sensation.

Objectives 13; I; 1; (1.7).   –   –   B

K(ii) Brain stem (Questions 942–945)

942 The red nucleus

(i) gives efferent fibres that excite mainly extensor
motor neurons.
(ii) is traversed by rootlets of the abducent nerve.
(iii) is at the same brain stem level as the inferior colliculus.
(iv) receives fibres from the cerebellum.

Objectives 13; I; 2; (2.1), (2.3), (2.4).
                  13; II; 3; (3.2).   *0.55*   81   D

943 The reticular formation

(i) is located in the brain stem and diencephalon.
(ii) forms part of a cortical activating system.
(iii) is composed of multiple short-chain neuronal
pathways.
(iv) is located in the cerebellum.

Objectives 13; I; 4; (4.4), (4.2).   –   –   A

944   The spinal nucleus of V

(i)   is continuous with the substantia gelatinosa.
(ii) receives terminations of fibres having cells of origin in the trigeminal ganglion.
(iii) is associated with pain and temperature sensation.
(iv) lies dorso-lateral to the nucleus ambiguus.

Objectives   13; I; 3; (3.2), (3.3), (3.4).   0.60   59   E
                                                      0.51   72

945   Cranial nerve efferent fibres arise from the

(i)   spinal trigeminal nucleus.
(ii) nucleus of the solitary tract.
(iii) substantia gelatinosa.
(iv) nucleus ambiguus.

Objectives   13; I; 3; (3.2), (3.3), (3.4).   0.58   58   D
                                                      0.60   67

K(iii) Sensory system (Questions 946–952)

946   Structures associated with the pathway of proprioception from the periphery to the central cortex include the

(i)   spinal ganglia.
(ii) posterior columns of spinal white matter.
(iii) medial lemniscus.
(iv) thalamus.

Objectives   13; I; 1; (1.4), (1.5).   13; II; 1; (1.1).   —   —   E

947   The medial lemniscus

(i)   carries signals from the ipsilateral side of the body.
(ii) terminates in the ipsilateral thalamus.

(iii) terminates in the ipsilateral post-central
gyrus.
(iv) consists of secondary afferent fibres.

Objectives   13; II; 1; (1.1).   *0.16*   67   C

948   Disturbed eye movement may result from damage
to the

(i)   medial longitudinal fasciculus.
(ii) inferior olive.
(iii) superior colliculus.
(iv) medial lemniscus.

Objectives   13; II; 2; (2.7), (2.8).   *0.44*   81   B
                                                  0.50   79

949   The retina projects directly to the

(i)   oculomotor nucleus.
(ii) superior colliculus.
(iii) visual cortex.
(iv) pretectal area.

Objectives   13; II; 2; (2.1), (2.3), (2.4), (2.6).   *0.29*   55   C
                                                      0.40   54

950   Auditory information is relayed in the

(i)   inferior olive.
(ii) nuclei of the trapezoid body.
(iii) lateral geniculate nucleus.
(iv) inferior colliculus.

Objectives   13; II; 1; (1.3).   *0.48*   87   C
                                                  0.40   85
                                                  0.50   80

951   Included in the connections of the medial longitudinal
fasciculus are the

(i)   vestibular nuclei.
(ii) nuclei of the 7th and 9th cranial nerves.

(iii) nucleus of the spinal part of the accessory nerve.
(iv) nucleus of the hypoglossal nerve.

Objectives   13; II; 2 (2.7).   *0.28*   43   **B**
                                                                 0.73   81

952   Some loss of cutaneous sensation on the opposite
      side of the body would result from cutting the

      (i)   medial lemniscus.
      (ii) internal capsule.
      (iii) lateral spinothalamic tract.
      (iv) lateral lemniscus.

      Objectives   13; II; 1; (1.1).   *0.63*   77   **A**

K(iv) Motor system and cerebellum (Questions 953–956)

953   Corticospinal fibres are found in the

      (i)   anterior funiculus of the spinal cord.
      (ii) cerebral peduncle.
      (iii) basilar part of the pons.
      (iv) dorsal portion of the mid-brain.

      Objectives   13; II; 3; (3.1).   *0.53*   70   **A**
                                                                 0.51   73
                                                                 0.58   74

954   Muscle tone in the lower limb is

      (i)   unaffected by cutting the lumbosacral dorsal
            roots.
      (ii) altered by the activity of the reticulospinal tracts.
      (iii) increased by damage to the ipsilateral cerebellar
            hemisphere.
      (iv) altered by the activity of the fasciculi proprii
            (intersegmental tracts).

      Objectives   13; I; 5; (5.3).   13; II; 4; (4.2).   *0.25*   57   **C**

955   Fibre tracts within the inferior cerebellar peduncle
include

   (i)   reticulocerebellar fibres.
   (ii) the anterior spinocerebellar tract.
   (iii) cerebellovestibular fibres.
   (iv) crossed fibres of the posterior spinocerebellar
         tract.

   Objectives   13; I; 5; (5.2).   *0.42*   24   B

956   A leg which has become paralysed as a result of
destruction of ventral horn cells usually has

   (i)   hypotonia.
   (ii) increased skin temperature.
   (iii) normal sensation.
   (iv) exaggerated reflexes.

   Objectives   13; II; 3; (3.4).   *0.55*   80   B

K(v) The circulation of the brain and cerebrospinal fluid (Questions 957–962)

957   The branches of the basilar artery include

   (i)   the posterior cerebral artery.
   (ii) branches to the cerebellum.
   (iii) branches to the pons.
   (iv) branches that supply the motor cortex.

   Objectives   13; III; 3; (3.1), (3.5).   *0.54*   92   A

958   The basilar artery

   (i)   enters the skull through the foramen magnum.
   (ii) is related to the ventral surface of the medulla.
   (iii) divides into two posterior cerebellar arteries.
   (iv) lies in the subarachnoid space.

   Objectives   13; III; 3; (3.1), (3.2).   —   —   D

959   An antero-posterior view of an internal carotid
      arteriogram shows

      (i)   the vertebral artery inferiorly.
      (ii) the anterior cerebral artery medially.
      (iii) division of the internal carotid into anterior
            and posterior cerebral arteries.
      (iv) the middle cerebral artery laterally.

      Objectives   13; III; 3; (3.6).   *0.72*   89   C

960   The fourth ventricle

      (i)   extends from medulla to mid-brain.
      (ii) is bounded laterally by middle cerebellar
            peduncles.
      (iii) opens rostrally into the third ventricle.
      (iv) has an uneven floor due to underlying cranial
            nerve nuclei.

      Objectives   13; III; 1; (1.1).   *0.47*   62   D

961   The falx cerebri

      (i)   is a fold of the meningeal layer of
            the dura mater.
      (ii) is attached to the tentorium cerebelli.
      (iii) is attached to the crista galli.
      (iv) has the straight sinus in its inferior free border.

      Objectives   13; III; 1; (1.8).   —   —   A

962   The tentorium cerebelli

      (i)   is attached to the petrous temporal bone.
      (ii) is attached to the wall of the transverse sinus.
      (iii) is in contact with the occipital lobe.
      (iv) has a free edge that is closely related to the
            mid-brain.

      Objectives   13; III; 1; (1.8).   *0.58*   79   E

K(vi) Cerebrum (Questions 963–970)

963   Extensive damage to the left frontal lobe

(i)   may result from occlusion of the basilar artery.
(ii) is likely to result in psychiatric disturbances.
(iii) usually results in loss of hearing.
(iv) produces a serious paralysis of the contralateral
      limbs.

Objectives   13; III; 2; (2.2), (2.3).   *0.24*   67   C
                                                0.37   80
                                                0.63   91

964   Psychiatric disturbance would be likely to follow
      extensive damage to the

(i)   globus pallidus.
(ii) septal area.
(iii) ventral part of the pons
(iv) amygdala.

Objectives   13; III; 4; (4.1), (4.3).   *0.63*   91   C

965   Pathways of the limbic system include

(i)   fornix.
(ii) stria terminalis.
(iii) cingulum.
(iv) corona radiata.

Objectives   13; III; 4; (4.2).   *0.32*   74   A

966   Connections of the hypothalamus include the

(i)   hippocampus.
(ii) cerebral cortex.
(iii) tegmentum of the mid-brain.
(iv) hypophysis cerebri.

Objectives   13; III; 4; (4.2), (4.4).   *0.37*   64   E

967   On the cerebral hemisphere the

(i)   area associated with somatic sensation is in the parietal lobe.
(ii) visual area is on the medial side of the occipital lobe.
(iii) auditory area is in the temporal lobe.
(iv) motor speech area is in the superior frontal gyrus.

Objectives   13; II; 1; (1.1), (1.3).   13; II; 2; (2.4).   *0.57*   67   A
             13; III; 2; (2.2).   *0.31*   61
                                                            0.30   65

968   The calcarine sulcus

(i)   lies on the inferior surface of the cerebrum.
(ii) separates the parietal from the occipital lobe.
(iii) is a sulcus of the parietal lobe.
(iv) lies within the primary visual area.

Objectives   13; I; 6; (6.1).   *0.52*   85   D
                                                            0.64   88

969   Fibres that link the right and left cerebral hemispheres cross in the

(i)   corpus callosum.
(ii) body of the fornix.
(iii) anterior commissure.
(iv) internal capsule.

Objectives   13; I; 6; (6.1).   13; III; 4; (4.2).   *0.27*   32   A
                                                            0.21   39

970   Functions of the corpus callosum include

(i)   connection of regions of the brain concerned with emotion and its expression.
(ii) integration of functions in the cerebral hemispheres.

(iii) transference in relation to cerebral functions which are not bilaterally represented.
(iv) transmission of corticospinal fibres.

Objectives   13; I; 6; (6.6).   13; III; 2; (2.2).   *0.23*   74   A

K(vii) Physiology (Questions 971–1000)

971   Glycine

(i)   is the main inhibitory transmitter in the mammalian spinal cord.
(ii) produces an increased permeability of the post-synaptic membrane to potassium or chloride ions.
(iii) is antagonized by strychnine.
(iv) is not involved in presynaptic inhibition.

Objectives   −   *0.38*   48   E

972   The caudate nucleus

(i)   receives a dopaminergic input from substantia nigra.
(ii) projects to the putamen.
(iii) is abnormal in Parkinson's disease.
(iv) projects to the spinal cord.

Objectives   13; II; 3; (3.1), (3.7).   *0.38*   76   A

973   The angular direction of sounds in space can be localized by using as cues:

(i)   loudness differences between the two ears.
(ii) conduction distance to the superior olivary nucleus.
(iii) time differences between the two ears.
(iv) the directionality of the external acoustic meatus.

Objectives   −   *0.35*   75   B

974 Horizontal spinning to the right is associated with

(i) inhibition of haircells in the ampulla of the left horizontal semicircular canal.
(ii) nystagmus (quick phase) to the right during rotation.
(iii) post-rotational nystagmus to the left.
(iv) past-pointing to the left.

Objectives —   0.37   68   A

975 The 'gate control' hypothesis of pain postulates that

(i) activation of large fibres excites an inhibitory interneurone.
(ii) large and small fibres converge on to one neurone in the spinal cord.
(iii) activation of small fibres inhibits an inhibitory interneurone.
(iv) transmission of impulses towards the cerebral cortex is reduced by activation of large fibre systems.

Objectives —   0.43   82   E

976 In the fovea

(i) there is a considerable convergence from bipolars to ganglion cells.
(ii) rhodopsin is the principal visual pigment.
(iii) absolute threshold is lowest.
(iv) blue cones are absent.

Objectives —   0.49   77   D

977 Gamma-aminobutyric acid

(i) is the main inhibitory transmitter in the spinal cord.
(ii) is not involved in presynaptic inhibition.

(iii) causes an increased permeability of the post-
synaptic membrane to potassium ions.

(iv) is antagonized by picrotoxin.

Objectives —   0.02   29   D

978 The basal ganglia

(i)   receive a dopaminergic input from the substantia
nigra.

(ii) project to the substantia nigra.

(iii) receive a cholinergic input from the cerebral
cortex.

(iv) project to the cerebral cortex.

Objectives   13; II; 3; (3.1).   0.37   56   A

979 Kittens reared to 3 months with both eyes covered

(i)   do not show a visual placing reaction.

(ii) do not show a blink reflex.

(iii) have an increased proportion of unresponsive
cells in the primary visual cortex.

(iv) have a changed distribution of binocularly
activated cells in the primary visual cortex.

Objectives —   0.34   35   A

980 The excitatory post-synaptic potential in a
mammalian motor neurone produced by
stimulation of Ia nerve fibres

(i)   is a depolarization lasting about 20 ms.

(ii) is triggered by the release of acetylcholine from
the afferent neurone.

(iii) is due to an increased permeability of the motor
neurone membrane to sodium and potassium
ions.

(iv) will lead to an action potential if its amplitude
exceeds about 1 mV.

Objectives —   0.40   50   B

981   Cells in S1 (primary somatosensory) cortex

    (i)   lie in the parietal lobe.
    (ii) have receptive fields on the contralateral half
        of the body.
    (iii) are organized somatotopically.
    (iv) are prıncipally excited by noxious stimuli.

Objectives   13; II; 1; (1.1).   *0.60*   92   **A**

982   Cone photoreceptors

    (i)   have their greatest density in peripheral retina.
    (ii) are activated principally in dim illumination.
    (iii) contain rhodopsin as their principal visual
        pigment.
    (iv) adapt before rods when a subject moves into a
        dark room.

Objectives   −   *0.56*   90   **D**

983   Mild hyperopia might not be detected because

    (i)   the far point is only slightly less than normal.
    (ii) the near point is only slightly greater than
        normal.
    (iii) the range of accommodation is greater than
        normal.
    (iv) by accommodating, the subject can focus on
        distant objects.

Objectives   −   *0.42*   65   **C**

984   Rapidly adapting mechanoreceptive fibres

    (i)   are absent from joints.
    (ii) are found in the posterior columns of the spinal
        cord.
    (iii) convey information about duration of the
        stimulus.
    (iv) signal vibration.

Objectives   13; II; 1.   *0.42*   65   **C**

985 The primate lateral geniculate nucleus

    (i)   receives direct input from the entire retina of
           the contralateral eye.
    (ii) contains relay cells with X- and Y-like receptive
           field properties.
    (iii) contains relay cells which could be excited
           through both eyes.
    (iv) receives direct input from the ipsilateral striate
           area.

Objectives   13; II; 2.   *0.30*   82   C

986 Glycine

    (i)   is an inhibitory transmitter in the mammalian
           central nervous system.
    (ii) increases membrane permeability to chloride ions.
    (iii) increases membrane permeability to potassium
           ions.
    (iv) is antagonized by strychnine.

Objectives   −   *0.47*   90   E

987 Gamma motor neurones

    (i)   innervate extrafusal muscle fibres.
    (ii) have their cell bodies in the ventral horn.
    (iii) cause increased discharge of group Ib nerve
           fibres.
    (iv) increase the sensitivity of the muscle spindle to
           stretch.

Objectives   13; II; 3; (3.3).   *0.59*   80   C

988 Pyramidal tract neurones

    (i)   activate motor neurones only via interneurones.
    (ii) cause specific movement of joints.
    (iii) do not activate specific motor neurones.

(iv) can discharge in the absence of skeletal muscle activation.

Objectives   13; II; 3; (3.1), (3.4).   *0.30*   34   D

989   The plantar response

(i)   includes dorsiflexion of the big toe in the adult.
(ii)   includes plantar flexion of the big toe in the infant.
(iii) becomes a plantar flexion of the big toe when descending motor pathways are injured in the adult.
(iv)   is a postural reflex.

Objectives   13; II; 4; (4.2), (4.4).   *0.33*   75   D

990   The impedance mismatch between air and cochlear fluid is overcome by

(i)   the ratio of areas of tympanic membrane to footplate of stapes.
(ii)   the leverage due to curvature of the tympanic membrane.
(iii) the lever action of the ossicles.
(iv)   the action of the stapedius muscle.

Objectives   −   *0.46*   65   A

991   The cells of the cuneate nucleus

(i)   have receptive fields which are only excitatory.
(ii)   are modality specific.
(iii) give rise to the fibres of the cuneate tract.
(iv)   are somatotopically organized.

Objectives   13; II; 2; (2.1).   *0.47*   90   C

992   Horizontal cells in the retina synapse with

(i)   rods.
(ii) cones.
(iii) bipolar cells.
(iv) amacrine cells.

Objectives   13; II; 2; (2.1).   *0.47*   90   A

993   Hyperpolarization of photoreceptors

(i)   is normally present in the dark.
(ii) is associated with decreased resistance of
      photoreceptor membrane.
(iii) inhibits horizontal cell activity.
(iv) is the response to light.

Objectives   −   *0.40*   64   D

994   The oculomotor nuclei

(i)   contain cells belonging to the somatic efferent
      columns.
(ii) are bounded laterally by the medial
      longitudinal fasciculus.
(iii) give origin to axons which emerge ventrally
      from the mid-brain.
(iv) give origin to all the somatic motor fibres of
      CN III.

Objectives   13; I; 3.   −   −   E

995   When red goggles are worn in daylight

(i)   dark adaptation occurs.
(ii) blue cones are active.
(iii) rods are not excited.
(iv) rod interference with cone adaptation is
      prevented.

Objectives   −   *0.37*   55   B

996   The retinal ganglion cells project directly to the

(i)   oculomotor nucleus.
(ii) superior colliculus.
(iii) visual cortex.
(iv) pretectal area.

Objectives   13; II; 2.   *0.33*   73   C

997   Cells in the inferior colliculus

(i)   display lateral inhibition.
(ii) give rise to the lateral lemniscus.
(iii) have tuning curves which are narrower than
      those of primary afferent fibres.
(iv) mediate visual reflexes.

Objectives   13; II; 1.   *0.41*   67   B

998   The basilar membrane

(i)   vibrates maximally to low frequency sounds at
      its apex.
(ii) separates the scala media from the scala tympani.
(iii) is tonotopically organized.
(iv) has a narrowly tuned resonance so that few hair
      cells are activated by any particular frequency.

Objectives   −   *0.45*   47   A

999   Centres involved in the control of micturition are
       found in the

(i)   hypothalamus.
(ii) spinal cord.
(iii) mid-brain.
(iv) cerebral cortex.

Objectives   13; III; 5.   *0.45*   54   E

1000 Cells in the olfactory bulb

  (i) are activated by horizontal cells.
  (ii) are inhibited by efferent fibres from the brain.
  (iii) are ciliated.
  (iv) project directly to the olfactory cortex.

  Objectives —   *0.41* 37   C